江苏大学英文教材基金资助出版
留学本科英文授课校级精品课程基金资助出版

CIVIL ENGINEERING SURVEYING
A GUIDE FOR EXPERIMENT AND PRACTICE

土木工程测量实验
与实习指导教程

（中英文双语版）

张秀丽（Zhang Xiuli）
韩　豫（Han Yu）
编著

江苏大学出版社
JIANGSU UNIVERSITY PRESS
镇江

图书在版编目(CIP)数据

土木工程测量实验与实习指导教程 = Civil Engineering Surveying：A Guide for Experiment and Practice：汉、英 / 张秀丽，韩豫编著. — 镇江：江苏大学出版社，2023.12
　　ISBN 978-7-5684-1954-3

Ⅰ. ①土… Ⅱ. ①张… ②韩… Ⅲ. ①土木工程－工程测量－实验－高等学校－教学参考资料－汉、英 Ⅳ. ①TU198-33

中国国家版本馆 CIP 数据核字(2023)第 251235 号

土木工程测量实验与实习指导教程
Civil Engineering Surveying：A Guide for Experiment and Practice

编　　著	张秀丽　韩　豫
责任编辑	郑晨晖
出版发行	江苏大学出版社
地　　址	江苏省镇江市京口区学府路 301 号(邮编：212013)
电　　话	0511-84446464(传真)
网　　址	http://press.ujs.edu.cn
排　　版	镇江市江东印刷有限责任公司
印　　刷	江苏凤凰数码印务有限公司
开　　本	787 mm×1 092 mm　1/16
印　　张	8.75
字　　数	238 千字
版　　次	2023 年 12 月第 1 版
印　　次	2023 年 12 月第 1 次印刷
书　　号	ISBN 978-7-5684-1954-3
定　　价	45.00 元

如有印装质量问题请与本社营销部联系(电话：0511-84440882)

Preface

Civil engineering technology and international reputation are getting higher and higher with the vigorous development of infrastructure in China in recent years. More and more foreign students who are attracted to China come here to study civil engineering. However, the construction of the matching English textbooks or bilingual textbooks, Chinese and English, lags behind. The related books are very scarce in the market. In view of this condition, this bilingual textbook, *Civil Engineering Surveying: A Guide for Experiment and Practice* ,written and reviewed by the author can not only fill this gap, but also serve the bilingual teaching of civil engineering and related majors in China. The purpose of this book is to strengthen practical teaching, train basic skills of engineering surveying, improve the abilities of instrument operation, draw and calculation, help students consolidate and deepen their understanding of theoretical knowledge, cultivate their abilities to analyze and solve problems, and develop a rigorous work style.

This book is a guide for civil engineering surveying experiments and internships. It is written by the author based on teaching reform and summaries of years of surveying practice teaching experience, combined with the current professional teaching plan, the content and requirements of theory curriculum.

The book is mainly divided into two parts. There are 14 civil engineering surveying experiments in the first part. It introduces the basic theoretical knowledge, experimental objectives, instruments, principles, steps, precautions, etc. in surveying experiments and practice systematically, and provides standard tables for recording, which can help students understand the experimental content and strengthen the experimental results better. The second part is the civil engineering surveying practice, which combines theoretical knowledge with individual experimental techniques for comprehensive training. The main content includes the purpose and nature, requirements for contents and results, arrangement, organization, grade evaluation, and precautions. Students can further understand and strengthen theoretical knowledge, cultivate problem-solving and hands-on abilities through the practice.

We referred to a large number of materials during the development of this book, and we would like to express our sincere gratitude to the authors of the referenced Chinese and foreign literature. Due to the limited level of editors, it is inevitable that there are deficiencies in the book. Readers are kindly requested to provide valuable feedback and suggestions.

前　言

近年来，随着国家基础建设的大力发展，我国的土木工程技术水平以及国际声望越来越高，慕名到我国学习土建类相关专业的留学生越来越多，然而与之相配套的全英文教材或者中英文双语教材的建设还比较滞后，市场上非常缺乏。鉴于此，作者编写了中英文双语版的《土木工程测量实验与实习指导教程》。这样既能解决教材建设滞后和市场缺乏的问题，也能为土建类相关专业的双语教学服务。本书旨在强化实践教学，通过实验与实习，使学生掌握工程测量的基本技能，提高对仪器的熟练运用能力，锻炼绘图和计算的能力，并巩固和深化对理论知识的理解，提高分析问题和解决问题的能力，养成严谨的工作作风。

本书为土木工程测量实验与实习指导用书，是作者基于多年的测量实践教学经验总结以及教学改革，结合目前专业教学计划和配套教材的内容与要求编写而成的。

本书主要分为两部分，第一部分是土木工程测量实验，共14个实验，系统地介绍了测量实验和实习中的基本理论知识、实验目的、实验仪器、实验原理、实验步骤、注意事项等，并给出了用于记录的标准表格，能够较好地帮助学生理解实验内容和巩固实验效果。第二部分是土木工程测量实习，是将理论教学与单项实验技术相结合进行综合训练的教学环节，主要内容包括实习的目的、性质，实习内容及要求，实习成果要求，实习时间安排，实习组织，实习成绩评定及注意事项等，通过实习可让学生进一步理解和巩固理论知识，培养解决问题的能力和动手能力。

本书在编写过程中参考了大量的资料，在此向参考过的中外文文献的作者表示诚挚的感谢。由于作者水平有限，书中难免存在不足之处，敬请读者提出宝贵意见和建议。

Contents
目 录

Requirements for Civil Engineering Surveying Experiment and Practice /001

Part I Civil Engineering Surveying Experiment /004

Experiment 1   Use of Automatic Levels   /004

Experiment 2   Technical Leveling   /009

Experiment 3   Fourth-Order Leveling   /012

Experiment 4   Testing and Adjusting of Levels   /015

Experiment 5   Use of Optical Theodolites   /020

Experiment 6   Horizontal Angles Observation   /025

Experiment 7   Vertical Angle Measurement and Index Error Adjustment   /028

Experiment 8   Testing and Adjusting of Theodolites   /032

Experiment 9   Steel Tape Measurement   /039

Experiment 10   Use of Total Stations   /043

Experiment 11   Understanding and Using of GPS   /051

Experiment 12   Staking out Axis and Elevation of Building   /056

Experiment 13   Staking out Circular Curve of Road   /060

Experiment 14   Unmanned Aerial Vehicle Topographic Mapping   /065

Prat II Civil Engineering Surveying Practice   /070

土木工程测量实验与实习要求　/073

第一部分　土木工程测量实验　/075

实验 1　自动安平水准仪的使用　/075

实验 2　工程水准测量　/079

实验 3　四等水准测量　/082

实验 4　水准仪的检验与校正　/085

实验 5　光学经纬仪的使用　/089

实验 6　水平角观测　/093

实验 7　竖直角测量及竖盘指标差检校　/096

实验 8　经纬仪的检验和校正　/099

实验 9　钢尺量距　/104

实验 10　全站仪的使用　/107

实验 11　GPS 的认识与使用　/113

实验 12　建筑物轴线测设和高程测设　/117

实验 13　道路圆曲线测设　/121

实验 14　无人机地形图测绘　/125

第二部分　土木工程测量实习　/129

参考文献　/132

Civil Engineering Surveying Experiment and Practice Requirements

Notes for surveying practice

(1) Everyone should make preparations before class, including previewing contents of relevant experiments in the textbook, understanding learning purpose and requirements of the experiments, familiarizing experiment steps, and preparing necessary forms and stationery.

(2) Everyone should observe class disciplines. Absent for no reason, attending class late or leaving early are prohibited.

(3) The tasks assigned by teachers should be fulfilled conscientiously.

(4) Practice should be conducted in a designated location, which can not be altered without authorization.

(5) During practice, instruments and tools should be taken good care of and measuring instruments should be strictly operated in accordance with the usage standards of them.

(6) When recording, the rules of recording measurement data must be strictly observed.

(7) Trees, flowers and crops should be cared for in the practice and should not be damaged arbitrarily.

Rules for recording measurement data

(1) The experimental records should be filled in the prescribed forms directly, and cannot be recorded on another paper and then be copied into the record sheet.

(2) Records and calculations must be written in H or 2H pencils, not in pens, ball-pens or other pens.

(3) All records should be clear written in the specified sheet, and the upper part of the sheet should be left with appropriate blank for error correction.

(4) The wrong number should be crossed out with a horizontal line, and the correct number should be written above the original number. It is forbidden to alter the original words by covering them or wipe them with an eraser.

(5) It is forbidden to change the number continuously, such as changing the observation data first and change its average then. In principle, the mantissa of observation data should not be changed, such as minutes and seconds of angle, centimeters and millimeters of level and

distance.

(6) The recorded number should be complete, the number 0 in them should not be omitted casually.

(7) When one person observes and the other records, the recorder should repeat the figures to the observer.

Rules for the use of measuring instruments

Instruments used in civil engineering surveying practice are mostly delicate and expensive. To ensure the safety of the instrument, extend its service life, and maintain its accuracy, the following requirements must be followed when using it.

(1) Optical instruments should be stored in strictly moisture-proof, dust-proof and shock-proof place. They should not be used in rainy days or in windy and sandy weathers.

(2) Instruments should try to avoid from being erected on traffic arteries, and there must be someone beside guarding them.

(3) After setting up the instrument, it is necessary to check whether its tripod leg screws and connecting screws are tightened.

(4) If an instrument would be moved in a short distance during use, its clamp screws should be released. For theodolite, its telescope should be placed vertically, the whole instrument should be held in front of chest, and its base part should be supported with one hand, the theodolite should not be carried on shoulder.

(5) When twisting the screw of an instrument, proper force should be exerted. When the clamp screw is not loosened, the alidade and the telescope should not be rotated.

(6) Do not sit on an instrument box while working. When an instrument is transported in box, it should be checked whether the buckle is fastened and the belt is durable.

(7) The work should be suspended and the instrument should be immediately stopped and inspected if any malfunction or abnormal sound is found during use. This situation should be immediately reported to the relevant laboratory staff.

(8) If the optical part of an instrument is contaminated with dust, it should be swept with a soft brush. It should not be wiped with neither dirty or rough cloth nor hand.

(9) The condition of the instrument, especially whether the accessories are lost or not, should be thoroughly checked after an instrument is used.

(10) The original position should be kept and the clamp screw should be released, when instrument is packed. Don't press hard if an instrument box cannot be tightly covered, check the position of the instrument is right or not. If not, adjust till the position is proper and then cover the instrument box.

(11) When using a steel tape measurement, do not pull it while coiling, and avoid

stepping on it or let it being pressed by a car. A steel tape must be wiped clean, oiled, and then drawn into the box after being used.

(12) The tape should be pulled with 1~2 turns rolled up, when measuring a distance, and the force should not be too strong to avoid damaging the connecting part.

(13) Range rods and leveling staffs should be clearly marked and non-curved. They should not be used to carry other things or left lying around.

Part Ⅰ Civil Engineering Surveying Experiment

Experiment 1 Use of Automatic Levels

Use of automatic levels

1.1 Purpose and requirement

(1) Understand the basic structure and performance of automatic level (DSZ3), know the name and function of its main components.

(2) Practice the placement and aiming operation of automatic level, and master the reading method and height difference calculation method.

1.2 Arrangement and equipment

(1) Experimental arrangement: This experiment can be arranged for 1 ~ 2 class hours. Each group consists of 3 persons, one person operates the instrument, one person sets up the leveling rod, and another person records data.

(2) Experimental equipment (per group): 1 DSZ3 level, 2 leveling rods, and 1 recording board. 2H pencil and calculator are self-provided.

1.3 Methodology and procedure

1.3.1 Understand the structure and working principle of an automatic level

Figure 1-1 shows the structural diagram of DSZ3 automatic level.

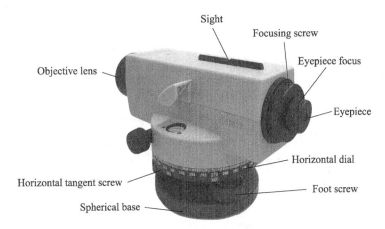

Figure 1-1 Structure of automatic level DSZ3

By using circular level, the compensating prism in an automatic level automatically level the instrument's collimation axis under action of gravity (completes precise leveling).

1.3.2 Placement of level and leveling

1. Placement of a tripod and connection of instrument

The point where instrument is placed is called the measuring station. Loosen the nuts on the tripod legs at the selected station first, adjust the tripod legs' height to get stability then, finally tighten the nuts again. Set the tripod, making the height of tripod's head a little lower than operator's shoulder. Make the head of the tripod as horizontal as possible, step on bottom of the tripod to make the tripod stable. Then take out the level from instrument box and put it on the head of the tripod, hold the level with one hand, rotate the connecting screw on the tripod into the pedestal of the level to fix it with the other hand, gently push the level to check whether it is really firmly connected.

2. leveling

The leveling of an automatic level is achieved by turning the foot screw to make the bubble center in round level. As shown in Figure 1-2a, when the bubble is not centered and located at position a, operator can simultaneously rotate the leveling screws ① and ② relatively to each other in the direction shown in the figure. The bubble's moving direction follows the left thumb's moving direction when turning the screws; if the bubble moves to position b, rotate another leveling screw ③ to center the bubble, as shown in Figure 1-2b. Repeat the above operation until the bubble is centered when rotating the telescope to any direction.

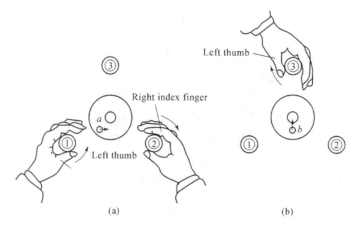

Figure 1-2 Leveling of automatic level

3. Aiming at a leveling rod

When conducting leveling measurements, the steps to aim at a leveling rod with a telescope are as follows:

(1) Eyepiece focusing. Point the telescope towards a bright background and rotate the eyepiece focus screw, making the cross hairs clear.

(2) Rough targeting. Rotate the alidade of the instrument to align the eyepiece, the sight, and the leveling rod in the same straight line.

(3) Objective lens focusing. Observe from a telescope, focus the objective lens to make the image of the level clear.

(4) Accurate sighting. Rotate the horizontal tagent screw so that the vertical line of the cross hair aligns with the center or edge of the leveling rod.

(5) Parallax elimination. The reading on the level rod changes if the eye moves slightly up and down near the eyepiece, this phenomenon is called parallax. This is because the image does not coincide with the cross hair plane. Parallax can cause significant reading errors, so it must be eliminated. The method to eliminate parallax is to carefully rotate the focusing screws of the eyepiece and the objective lens until the ruler image coincides with the cross hair plane.

4. Reading

After sighting the leveling rod, operator should level the instrument first, and then take the value gotten from middle line of the crosshairs as the reading. Before reading, be sure to press the compensator(operate according to instrument model). Read from top to bottom (inverting telescope), from small to large when getting the value on the level rod, and estimate each reading to "mm", and get four digits. It should be noted that the details of different models of leveling rods may vary and should be evaluated based on the actual situation.

In summary, the basic operating procedures of a level can be simply summarized as follows: placing the instrument-leveling-sighting-reading.

1.3.3 Leveling records

After each person practices placement of a level, the two vertical leveling rods should be sighted and read. Measurement data should be recorded in Table 1-1. Height difference between setting spots of the two leveling rods should be calculated. After all measuring stations' leveling tasks are finished, submit Table 1-1 as a result of the whole experiment.

1.4 Precaution

(1) After placing the instrument on the tripod, the connecting screw must be snugly screwed to ensure tight connection and prevent the instrument from falling off the tripod and damaging.

(2) Before reading, parallax must be eliminated when sighting at a target.

(3) The reading gotten from the leveling rod must have four digits: m, dm, cm, mm. The first digit should be zero if the reading is less than 1 m, the corresponding digits should also be filled with zero if the reading is an integral decimeter or centimeter.

Table 1-1 Leveling reading exercise

Class: _____ Group: _____ Date: _____ Observer: _____
Instrument tag number: _____ Recorder: _____

Station	Mark	Reading/mm		Height difference/m	Average height difference/m	Note
		BS	FS			

Note: BS refers to backsight; FS refers to foresight.

Experiment 2 Technical Leveling

2.1 Purpose and requirement

(1) Master how to select measuring stations and turning points, how to hold a level rod, as well as how to operate instruments.

(2) Master surveying, recording, adjusting of height difference misclosure and elevation calculation of technical leveling(observed by two different instrument heights).

2.2 Arrangement and equipment

(1) Experimental arrangement: 2 class hours for the experiment, 4 people in each group. The division of labor is as follows: one person operates an instrument, one person records data and two persons set up two leveling rods.

(2) Experimental equipment(per group): 1 DSZ3 level, 2 level rods, 2 steel turning pins and 1 recording board. 2H pencil and calculator are self-provided.

2.3 Methodology and procedure

2.3.1 Observation method by two different instrument heights

Two different instrument heights method, also known as the variable instrument height method, is to check the height difference obtained from two different instrument heights on the same measuring station. That means after measuring the first height difference, change the height of the instrument (the height difference of the instrument should be larger than 10 cm) and measure again. The difference between the two results cannot exceed the allowable value (±5 mm for ordinary leveling), and the average is taken as the result.

2.3.2 Procedures

(1) Starting from a benchmark on an experimental area, build a leveling loop, as shown in Figure 2-1, 4~5 stations can be set along this loop, and the sight length should be about 50 m. Positions of fixed features with protruding points or marked points can be chosen as set positions of the leveling rods, if the

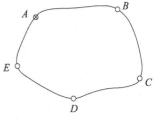

Figure 2-1 Leveling loop

ground is unfirm, a turning point plate can be placed to prevent the leveling rod from settling.

(2) As shown in Figure 2-1, set up a leveling instrument approximately halfway between the starting point A (a benchmark) and the first turning point B (backsight and foresight

distances are balanced by range estimation or pacing) and roughly level it. An observer should observe in the following order:

① Sight a leveling rod on point A(backsight point), then finely level and read;

② Sight a leveling rod on point B(foresight point), then read;

③ Change the height of the instrument more than 10 cm and reset the level;

④ Sight the leveling rod on point B(foresight point), then finely level and read;

⑤ Sight the leveling rod on point A(backsight point), then read.

(3) When recording data, for each read obtained by the observer, the recorder should repeat the value and record it on site. After the backsight reading and foresight reading are completed, the height difference should be calculated and checked at once.

(4) Set stations in turn and measure the hight difference with the same method until back to the starting point.

(5) The leveling loop should be checked, the sum of foresight readings, backsight readings and height difference should be calculated after the survey is completed. It indicates that the calculation is correct if the sum of the backsight readings minus the sum of the foresight readings equals the sum of the height difference.

2.4 Precaution

(1) The leveling rod must be held vertical when sighting and reading by a level. Observer can detect the left or right inclination of the leveling rod according to relative position of vertical hair and leveling rod. The rod setting person can detect whether the level rod is tiltered or not in the observation direction by the "L" shape, which is formed by the observation direction and the extended arm with the rod in hand.

(2) The difference between the two results measured by the two instrument heights method should not be greater than 5 mm for each measuring station, otherwise, the station should be resurveyed.

(3) Each station can only be moved after passing the check. If there are turning point plate under the leveling rods on the backsight point and the foresight point, then the leveling rods cannot be moved until the level is moved. It indicates that it has passed the check when the instrument is moved, and the rod setting person on the backsight point can carry the rod and turning point plate to next point; the rod setting person on the foresight point still can not move the turning point plate, but turn over the level rod, changing its direction from foresight to backsight.

(4) The allowable height difference misclosure of closed level route is $\pm 6\sqrt{n}$ mm for hilly area(n is the number of measuring stations), $\pm 20\sqrt{L}$ mm for flatland(L is the length of level route, its unit is km).

(5) Record the data in the Table 2-1, it should be submitted as a result of this experiment after completing the corresponding calculation.

Table 2-1 Engineering leveling record

Class: _____ Group: _____ Date: _____ Observer: _____
Instrument tag number: _____ Recorder: _____

Station	Mark	Reading/mm		Height difference/ m	Average height difference/ m	Adjusted height difference/ m	Elevation/ m	Note
		BS	FS					
Verifying calculation $\sum$								

Experiment 3　Fourth-Order Leveling

3.1　Purpose and requirement

(1) Master the observing and recording method of the fourth-order leveling.

(2) Get familiar with the organization and general regulations of the precision leveling, the main technical index of the fourth-order leveling, and master the checking methods for station and leveling route.

3.2　Arrangement and equipment

(1) Experimental arrangement: 2 class hours for the experiment, 2~3 people in each group.

(2) Experimental equipment(per group): 1 DSZ3 level, 1 tripod, 2 double-sided level rods, 2 turning point plate and 1 recording board. Calculator, 2H pencil, knife and paper for computation are self-provided.

3.3　Methodolgy and procedure

(1) Build a level loop or a connecting traverse, 4 or 6 stations can be set up along this route.

(2) Observation steps of fourth-order leveling (take DSZ3 level as an example) are as following:

① Roughly leveling an instrument with the circular vial.

② Sight to the BS, use the black side of the leveling rod, and read the values of the upper, lower and middle line after the instrument is leveled.

③ Turn over the level rod on the BS point, then read the value of the middle line on the red side of the level rod.

④ Sight to the FS, use the black side of the level rod, and read values of the upper, lower and middle line after the instrument is leveled.

⑤ Turn over the level rod on the FS point, then read the value of the middle line on the red side of the level rod.

(3) Fill in relevant data according to the requirements of fourth-order leveling observation and record, compute immediately after recording, and move the station only after the calculating results conform to the limit errors. In the computation, backsight distance=100×(BS upper line reading−BS lower line reading), foresight distance=100×(FS upper line reading−FS lower line reading).

(4) Set up stations in proper sequence and survey the other stations with the same method.

(5) After the field work is done, the misclosure, the adjustment of each station and the elevation of each control point should be computated.

3.4 Specification for fourth-order leveling

The specification for fourth-order leveling is shown is Table 3-1.

Table 3-1 Specification for fourth-order leveling

Level	Sight distance/m	Difference in sight distance/m	Accumulative difference in sight distance/m	Reading difference on two sides/mm	Height difference on two sides/mm
4	100	5.0	10.0	3.0	5.0

3.5 Checking

The verifying formulas of height difference are as following:

Black side: $\sum$ Backsight reading $-\sum$ Foresight reading $=\sum$ (Backsight reading $-$ Foresight reading)

Red side: $\sum$ Backsight reading $-\sum$ Foresight reading $=\sum$ (Backsight reading $-$ Foresight reading)

The verifying formula of the difference in sight distance is as following:

$\sum$ Backsight distance $-\sum$ Foresight distance $=$ Cumulative difference in sight distance at the end of the page $-$ Cumulative difference in sight distance at the end of last page.

3.6 Precaution

(1) The technical requirements for fourth-order leveling are strict, and the accuracy requirements for fourth-order leveling are high. The key lies in: backsight and foresight distances should be balanced (within the limit error); change from BS to FS (or vice versa), the instrument cannot be leveled again; the level rod should be vertical, it is better to use a level rod with a circular level.

(2) The records should be neat and clear. If there are mistakes in reading and record, data inolved must be resurveyed, and it is forbidden to change the records.

(3) The station cannot be moved until the records and calculations of this station are checked and corrected.

(4) Record the data in the Table 3-2, it should be submitted as a result of this experiment after completing the corresponding calculation.

Class: _____ Group: _____ Date: _____ Observer: _____ Instrument tag number: _____ Recorder: _____

Table 3-2 Fourth-order leveling

No.	Mark	Readings					Middle line at BS/m	Middle line at FS/m	Height difference/m	Adjustment of height difference/m	Adjusted height difference/m	Elevation/m
		Upper line/m	Lower line/m	Sight distance $(UL-LL) \times 100$/m	Difference in sight distance/m	Accumulate difference in sight distance/m						
								—				
							—					
								—				
							—					
								—				
							—					
								—				
							—					
								—				
							—					
								—				
							—					
Verifying calculation $\sum$		—	—		—							

Note: "——" refers to no data should be filled in the table.

Experiment 4 Testing and Adjusting of Levels

4.1 Purpose and requirement

(1) Understand the conditions that should be met between the axes of the level.

(2) Master the testing and adjusting methods of DSZ3 level.

4.2 Arrangement and equipment

(1) Experimental arrangement: 2 class hours for the experiment, 4 people in each group. One person operates the instrument for testing and adjusting, one person records data and two persons hold the level rods.

(2) Experimental equipment(per group): 1 DSZ3 automatic level, 2 leveling rods, 2 steel turning point plates, 1 small screwdriver, 1 adjusting pin and 1 recording board.

4.3 Methodology and procedure

4.3.1 Testing and adjusting principle of level

The axes and their positions of a DSZ3 automatic level are shown in Figure 4-1.

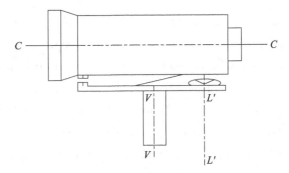

CC—collimation axis; VV—vertical axis; $L'L'$—axis of circular vial.

Figure 4-1 Schematic diagram of the axes and their positions of the DSZ3 automatic level

During leveling, a level must provide a horizontal line of sight. The line of sight is horizontal or not can be detected by the circular vial. Therefore, the collimation axis (CC) must be perpendicular to the axis ($L'L'$) of circular vial, which is the main condition that a level should meet. In addition, a level should also meet the following two conditions: ① the axis of circular vial is parallel to the vertical axis ($L'L' \parallel VV$); ② the horizontal crosshair is perpendicular to the vertical axis.

4.3.2 Testing and adjusting of level

1. Testing and adjusting of the circular vial axis $L'L'$

Rotate the three leveling screws to make the bubble stay in the center of the circular vial. It indicates that the condition of $L'L' // VV$ is met if the bubble is still in the center after the collimator is turned about 180° around its vertical axis, otherwise correction is required. The correction method: First loosen the fixing screw at the bottom center of the round level slightly, second turn the adjustment nuts of the circular vial to move the bubble half-way back to the centered position, then screw the foot screw of the circular vial to move the bubble back to center. Repeat the test until the bubble remains centered regardless of which direction the level is aimed at. Finally, tighten the fixing screw.

2. Testing and adjusting of the crosshair (see Figure 4-2)

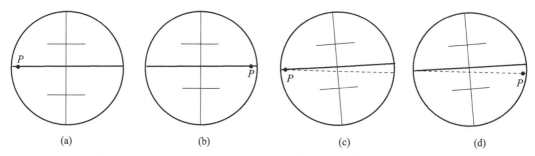

Figure 4-2 Testing and adjusting of the crosshair

(1) Testing method: Set up an instrument, level it. Sight a sharply defined point P with one end of the horizontal crosshair, and turn the horizontal tangent screw, if P always moves on the horizontal crosshair, as shown in Figures 4-2a and b, it indicates that the horizontal crosshair is perpendicular to the vertical axis of the instrument. If the moving track of P deviates from the horizontal crosshair, as shown in Figures 4-2c and d, it indicates that the horizontal crosshair is not perpendicular to the vertical axis of the instrument and it should be adjusted.

(2) Adjusting method: Unscrew the protective cover of the crosshair partition board, loosen the screws fixing the crosshair partition board from outside, finely rotate the reticle until point P always moves on the horizontal crosshair when moving the level horizontally. The screws should then be carefully tightened in their final position.

3. Testing and adjusting of collimation axis (see Figure 4-3)

Stake out four points equally spaced, make sure the distance between every two adjacent points is 40 meters as shown in Figure 4-3. Then drive wooden stakes or place turning point plates at points 1 and 2, hold level rods at points A and B. First a level is set up at point 1 and leveled, and then the rod readings r_A at A, and r_B at B are taken as shown in Figure 4-3a. Next the instrument is moved to point 2 as shown in Figure 4-3b. Readings (r'_A) at A and (r'_B) at B

are then taken after the instrument is leveled.

As illustrated in Figure 4-3a, the longer visual distance is 2 times longer than that of the shorter visual distance. Assuming that the reading error of the leveling rod at A caused by non-horizontal line of sight is e, then the reading error of the leveling rod at B is $2e$. Theoretically, the surveyed height differences between A and B should be the same, which gives

$$(r_B - 2e) - (r_A - e) = (r'_B - e) - (r'_A - 2e) \tag{4-1}$$

So

$$e = \frac{r_B - r_A - r'_B + r'_A}{2} \tag{4-2}$$

The formula for calculating the angle i between the collimation axis and the horizontal line is as follow (Note: 1 rad = 206 265"):

$$i = \frac{e}{D} \times 206\,265'' \tag{4-3}$$

Here, D means the horizontal distance between points A and B. The level needs not to be adjusted, if $i < 20''$, otherwise the instrument needs correction.

To make the axis of level tube parallel to the collimation axis, the calibration method is: remove the cover of the reticle, use the calibration needle to move the upper or lower calibration screws of the crosshair, and move the horizontal hair up or down until the required reading($r_A - e$) is obtained when sighting to the rod at A. Several trials may be necessary to achieve the exact setting during calibration.

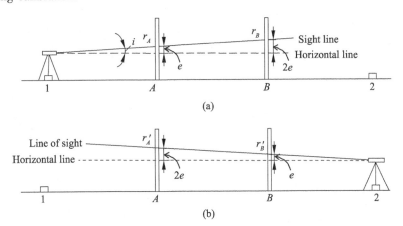

Figure 4-3 Testing and adjusting of collimation axis

4. Testing of compensating prism function

Sight to a level rod and get the reading. Tap the tripod foot with your hand to check the vibration of crosshair. It indicates that the function of the compensating prism is normal if the crosshair stabilizes quickly and its horizontal crosshair still aims at the original reading.

4.4 Precautions

(1) The testing and adjusting must be carried out in the order specified in the instruction, and cannot be reversed arbitrarily.

(2) The scope of one-time tightening should be small when turning the calibration screw, the correction screw should be slightly tightened after adjustment.

(3) The testing and adjusting records in Table 4-1 should be completed and submitted as the experimental results.

Table 4-1 Testing and Adjusting of level

Class: _____ Group: _____ Date: _____ Observer: _____

Instrument tag number: _____ Recorder: _____

Test items	Test procedure	
	Diagram	Observation data and description
Circular vial axis // Vertical axis		
Horizontal crosshair ⊥ Vertical axis		
Collimation axis is horizontal		$r_A =$ $r'_A =$ $r_B =$ $r'_B =$
		$e =$
		$i =$

Experiment 5 Use of Optical Theodolites

Use of theodolites

5.1 Purpose and requirement

(1) Understand the basic structure of an optical theodolite and functions and names of the main components.

(2) Master the basic operation methods of an optical theodolite—centering, leveling, sighting and reading.

5.2 Arrangement and equipment

(1) Experimental arrangement: 2 class hours for the experiment, 4 people in each group, team members take turns to use the instrument and record the data.

(2) Experimental equipment (per group): 1 optical theodolite, 1 recording board, 2 chaining pins.

(3) Several targets are placed for each group to practice sighting.

5.3 Methodology and procedure

5.3.1 Structure of an optical theodolite

Figure 5-1 illustrates the structure of an optical theodolite.

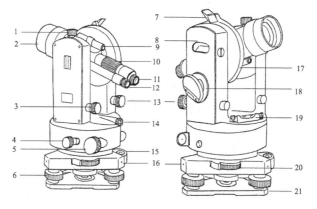

1—Telescope clamp screw; 2—Telescope objective lens; 3—Telescope tangent screw; 4—Horizontal clamp screw; 5—Horizontal tangent screw; 6—Foot screw; 7—Vertical circle level tube observation mirror; 8—Vertical circle level tube; 9—Collimator; 10—Objective lens focus ring; 11—Eyepiece; 12—Objective lens for reading; 13—Tangent screw of vertical circle level tube; 14—Optical plummet; 15—Circular level bubble; 16—Tribrach; 17—Vertical circle; 18—Dial illumination mirror; 19—Alidade tubular bubble; 20—Horizontal dial position change screw; 21—Base plate.

Figure 5-1 Structure of an optical theodolite

5.3.2 Setting up an optical theodolite

(1) Set up a tripod over a ground mark. The tripod legs are extended to suit the height of an observer. Take a few steps back from the tripod, the center of the tripod head is checked to see if it is plumb above the ground mark. The tripod head should be as horizontal as possible.

(2) An optical theodolite is taken out from the container, and being attached securely to the tripod head.

5.3.3 Basic operations of optical theodolite

1. Centering

(1) Rough centering by a plumb bob

Hang a plumb bob under the connecting screw, adjust the length of the plumb bob so that it can be a little bit higher than the ground mark, and then center it. If the plumb bob deviates a lot from the ground mark, move the legs of the tripod to roughly align the plumb bob with the ground mark, and then firmly insert the legs of the tripod into the ground. Finally, slightly loosen the center connecting screw and move the instrument on the tripod head slightly to center the plumb bob above the ground mark exactly. The error in the rough centering of the plumb bob should be within $0 \sim 5$ mm.

(2) Precise centering by optical plummet

The observer slightly looses the tribrach screw, translates the instrument on the tripod slowly while observing the optical plummet until the plummet's pointing device and the ground mark coincide. Since the top of the tripod may not be exactly leveled, the instrument should be translated (do not rotate it). If it is rotated while being translated, the sight line of the optical plummet will no longer be vertical. Both the leveling and centering must be checked after translation.

2. Leveling

The purpose of leveling theodolite is to make the horizontal circle of the theodolite in a horizontal plane. The procedures of leveling theodolite consists of rough leveling and precise leveling.

Rough leveling: Center the circular bubble by adjusting the lengths of the tripod extension legs. Refer to the procedures in experiment 1.

Precise leveling: Center the plate level bubble by turning the level-screws (Figure 5-2), the procedures are as follows:

① Rotate the theodolite to place the axis of the level vial parallel to the line through any two foot-screws. These two foot-screws are turned until the plate level bubble is brought to the center of its run. The foot-screws should be turned in opposite directions simultaneously, the bubble moves in the same direction of the left thumb when the foot-screws are turned (Figure 5-2a).

② Rotate the theodolite 90°, then turn the third foot-screw to centralize the plate level bubble (Figure 5-2b).

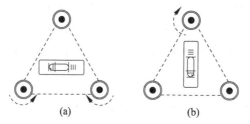

Figure 5-2 Precise leveling of the theodolite

Repeat the above steps until leveling and centering always meet requirements no matter which direction the theodolite is turned to.

3. Sighting

Loosen the horizontal clamp screw of the alidade, sight the object (chaining pin or target) with the collimator, then fix the clamp screws in horizontal and vertical directions. Focus the eyepiece to make the crosshair clear, focus the objective lens to make the target image clear, and eliminate the parallax (same as parallax elimination operation level). Rotate the telescope's vertical tangent screw to make the target image in proper height, rotate the horizontal tangent screw so that the target image is bisected by a single longitudinal wire of the crosshair or sandwiched in the center by two longitudinal wires to complete sighting.

4. Reading

First adjust the mirror located on one standard to reflect light into the instrument and properly brighten the circles; then focus the eyepiece to make the division of the dial clear. The reading system of the optical theodolite is shown in Figure 5-3. "Horizontal" or "H" is the reading of the horizontal dial; "Vertical" or "V" is the reading of the vertical dial. The micrometer reading of the dial is estimated to 0.1′ and converted into seconds. The horizontal dial in the figure reads 115°58′00″ and the vertical dial reads 79°04′30″.

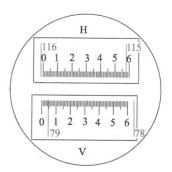

Figure 5-3 Reading system for optical theodolite

5. Other exercises

(1) Practice observing an angle in direct mode and reverse mode: loosen the horizontal and vertical clamp screws, rotate the telescope 180° around the horizontal axis, practice sighting at a target and reading the value.

(2) Practice changing the position of the horizontal dial: clamping the horizontal lock, open the protective cover, rotate changeable handwheel of the horizontal dial, observe the

changes in the horizontal dial reading from the reading eyepiece that adjacent to the telescope eyepiece, and try to set it with an integer reading, such as 0°00′00″, 90°00′00″, etc., then cover it.

5.4 Precaution

(1) The tripod head should be almost horizontal when centering a theodolite, otherwise it will cause difficulties in instrument leveling.

(2) The plate level bubble is centered or not in all directions should be checked, and the deviation of the bubble should be within the specified range when leveling the theodolite.

(3) Parallax must be eliminated when aiming at a target.

(4) Complete the Table 5-1 and submit it as the experimental result.

Table 5-1 Numerical reading exercise of horizontal dial

Class: _____ Group: _____ Date: _____ Observer: _____
Instrument tag number: _____ Recorder: _____

Station	Object	Vertical dial position	Horizontal dial reading °	′	″	Note
		L				
		R				
		L				
		R				
		L				
		R				
		L				
		R				
		L				
		R				
		L				
		R				

Experiment 6 Horizontal Angle Observation

6.1 Purpose and requirement

Master the operation, recording and computation methods of horizontal angle observation with theodolite through the direct and reverse mode.

6.2 Arrangement and equipment

(1) Experimental arrangement: 2 class hours for the experiment, 3 people in each group, group members take turns to use the instrument and record data.

(2) Experimental equipment (per group): 1 optical theodolite, 1 recording board, 1 or 2 chaining pins.

6.3 Methodology and procedure

As shown in Figure 6-1, it is assumed that a theodolite is set up over station B, observation targets are set at points A and C, the method and steps of measuring the horizontal angle β are as follow:

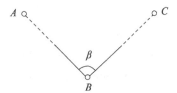

Figure 6-1 Observing horizontal angle by survey rounds

1. The first half round on face left

(1) Set the theodolite in face left (FL) position, rotate the changable handwheel of the horizontal dial, the horizontal circle is set to zero or near zero. Then set the screw on the horizontal dial to separate the aiming part from the horizontal dial (depending on the instrument).

(2) Loosen the horizontal and vertical clamp screws, sight to target A by sighting device, then fix both clamp screws. The vertical hair is accurately sighted onto the target A (usually at the bottom of target) using the horizontal and vertical tangent screws on the telescope. Attention must be paid to check and remove parallax. Finally take the reading "A_1" on the horizontal dial.

(3) Turn the alidade in a clockwise manner and sight to the target C, record the reading C_1 on the horizontal circle. Therefore, the horizontal angle β_1 between the targets A and C is

$\beta_1 = C_1 - A_1$.

2. The second half round on face right

The telescope is "plunged" (rotated 180° about the horizontal axis) so that the theodolite is set face right (FR). Sight to target C, take the reading C_2, then turn the telescope in an anticlockwise manner to aim at target A, and take the reading A_2. The difference between the two readings is the second half round angle β_2 ($\beta_2 = C_2 - A_2$), the first half round plus the second half round is one round, if the difference between β_1 and β_2 is not more than 42″, taking the average of β_1 and β_2 as the result of this horizontal angle [$\beta = (\beta_1 + \beta_2)/2$] for one round.

Take many repetitions if the accuracy of angle measurement is required to be high. To minimize the error caused by unequal disproportion of the horizontal dial, after each round, the dial reading should be adjusted by 180°/n according to its round time n. Example, if two survey rounds are needed, when starting the second set of angle, the observer should sight to A and set the horizontal dial's reading to 90°.

At least two rounds should be taken at each station in order to detect errors when the angles are computed. Since each round is observed independently, both rounds must be computed and checked, the instrument and tripod can be moved only when the error of the two rounds are acceptable. You'd better use the same point on the vertical hair whenever you aiming at the target in order to reduce the collimation error.

6.4 Precaution

(1) Plummet error should be less than 2 mm when placing theodolite.

(2) Aim at the bottom of a target as much as possible to reduce the error caused by target inclination when aiming at the target.

(3) The instrument should be releveled and this round should be remeasured if the tube-type level vial deviation exceeds 2 grids during observation.

(4) Each person should observe horizontal angle at least one round independently, fill the data in Table 6-1, and submit it as the results of this experiment.

Table 6-1 Horizontal angle survey (by survey rounds)

Class: _____ Group: _____ Date: _____ Observer: _____
Instrument tag number: _____ Recorder: _____

Station	Station sighted	Vertical dial position	Horizontal circle reading/ (° ′ ″)	Horizontal angle/(° ′ ″) Half round	Horizontal angle/(° ′ ″) One round	Note
		L				
		R				
		L				
		R				
		L				
		R				
		L				
		R				
		L				
		R				
		L				
		R				

Draw sketch here

Experiment 7 Vertical Angle Measurement and Index Error Adjustment

7.1 Purpose and requirement

(1) Understand the component, marked form of vertical circle, the relationship between vertical circle index error and plate bubble of theodolite.

(2) Master the measuring methods of vertical angle.

(3) Master the testing and adjusting methods of index error.

7.2 Arrangement and equipment

(1) Experimental arrangement: 2 class hours for the experiment, 2 people in each group.

(2) Experimental equipment(per group): 1 theodolite, 1 recording board.

(3) An instructor places several targets for each group to sighting.

7.3 Methodology and procedure

7.3.1 Vertical angle measurement

(1) Place an instrument over a designated station, center and level it, rotate the telescope, observe changes in readings of vertical circle from the eyepiece, confirm the marked form of the vertical circle, write the computing formulas of vertical angle and vertical circle index error in a record sheet.

(2) Select a target. Set the theodolite in face left and sight to the target (make the middle horizontal crosshair sight to the top of the target precisely). Turn the vertical tangent screw until the altitude bubble is brought to the middle of its run, then get the reading of vertical circle L and calculate the first half round vertical angle α_{left} ($\alpha_{left} = 90° - L$).

(3) Rotate the telescope so that the theodolite is now in face right position to start the other half round. Then get the reading of vertical circle R, the second half round vertical angle α_{right} can be calculated ($\alpha_{right} = R - 270°$).

(4) Calculate the index error $\overline{\chi}$ and the vertical angle of one round by the following formula.

$$\overline{\chi} = \frac{1}{2}(\alpha_{right} - \alpha_{left})$$
$$\alpha = \frac{1}{2}(\alpha_{left} + \alpha_{right})$$
(7-1)

(5) Each person should observe at least two rounds to the same target or one round for two different targets. The index error is a constant for a particular instrument. Therefore, the difference between the index errors measured each time should not be greater than 20″.

7.3.2 Vertical index error testing and adjustment

Check whether the difference of index error measured in each round is out of limit, eliminate outliers, and take the average as the vertical index $\overline{X}$ error of the instrument. If the absolute value of $\overline{X}$ is greater than 60″, the instrument needs to be adjusted.

The aim of the adjustment is to ensure that when the line of sight is horizontal and the plate bubble is central, the vertical circle reads 90° or some multiple of 90° (depends on the type of instrument).

1. Testing of vertical index error

Assume the circle graduation increasing in clockwise direction and the vertical circle reading is 90° when the theodolite is set in FL position and the line of sight is horizontal. The testing procedures are as follow:

(1) Set a theodolite in face left position, and level the instrument carefully.

(2) Make the horizontal hair of the telescope sight to a fine point precisely.

(3) Adjust the bubble to the middle of its run and take a vertical circle reading L.

(4) Turn the telescope, set the theodolite in face right position, and repeat step (2). Adjust the bubble again and take a vertical circle reading R.

For example, when checking the readings, FL vertical circle reading $L = 78°18'18''$. FR vertical circle reading $R = 281°42'00''$. According to formula (7-1), we obtain

$$\overline{X} = [(281°42'00'' - 270°) - (90° - 78°18'18'')]/2 = 9''$$

2. Adjustment of vertical index error

Set the theodolite in FR position, sight the line of sight to the fine point.

(1) When the theodolite in FR position, correct reading on vertical circle should be $(R-\overline{X})$, $(R-\overline{X}) = 281°42'00'' - 09'' = 281°41'51''$. This reading is set by rotating the altitude bubble leveling screw, which causes the altitude bubble to move off center.

(2) Adjust the altitude bubble capstan adjusting screws to recentralize the altitude bubble.

Note: For a theodolite with an automatic vertical index, the manufacturer's handbook should be consulted for the correct adjustment procedure.

7.4 Precaution

(1) Focus an eyepiece to make the crosshair clear and eliminate the parallax when sighting a target, try to sight an object by the crosshair's intersection, and make the bubble center before

each time reading.

(2) Pay attention to the positive and negative signs of measured vertical angles.

(3) Fill the data in Table 7-1, and submit it as the results of the experiment.

Table 7-1 Vertical angle measurement

Class: _____ Group: _____ Date: _____ Observer: _____

Instrument tag number: _____ Recorder: _____

Station	Station sighted	Position	Reading			Vertical angle						Index error
						Half round			One round			
			°	′	″	°	′	″	°	′	″	″
		FL										
		FR										
		FL										
		FR										
		FL										
		FR										
		FL										
		FR										
		FL										
		FR										
		FL										
		FR										
		FL										
		FR										
		FL										
		FR										
		FL										
		FR										
		FL										
		FR										
		FL										
		FR										
		FL										
		FR										
Computing formula of vertical angle and index error												

Experiment 8 Testing and Adjusting of Theodolites

8.1 Purpose and requirement

(1) Study the geometric relationships that the main axes of theodolite should meet.

(2) Master the basic method of theodolite testing and adjustment.

8.2 Arrangement and equipment

(1) Experimental arrangement: 2 class hours for the experiment, 3 people in each group.

(2) Experimental equipment (per group): 1 theodolite, 3 chaining pins, 1 leveling rod, 1 recording board and 1 adjusting pin.

8.3 Experimental principle

The instrument should be placed over the station exactly, the vertical axis and the control point should be on the same plumb line in order to measure horizontal or vertical angle correctly, It can form a vertical plane when the telescope rotates around the horizontal axis. The reading on the vertical dial should be 90° or 270° when the line of sight is horizontal. In order to meet the requirements, the instrument should meet the following requirements:

(1) The axis of the plate bubble should be perpendicular to the vertical axis;

(2) The vertical crosshair should be perpendicular to the horizontal axis;

(3) The collimation axis of the telescope should be perpendicular to the horizontal axis;

(4) The horizontal axis should be perpendicular to the vertical axis;

(5) The sight line of the optical plummet should coincide with the rotating centerline of the vertical axis;

(6) The vertical circle index should be in correct position.

This test is to check whether the instrument meets the above geometric relationships.

8.4 Methodology and procedure

8.4.1 Testing and adjusting of the tubular bubble axis perpendicular to the vertical axis

1. Testing requirement

(1) Everyone in the group should complete this test independently and make records.

(2) The twice testing results determine whether the instrument meets the geometric relationships. If the bubble offset in the two tests is less than half a grid, the relationship is

basically satisfied.

2. Testing steps

Level the instrument, make the plate bubble parallel to a pair of foot screws, turn the two foot screws to center the plate bubble in its run precisely, and then rotate the alidade 180°. It means that the axis of the plate bubble is perpendicular to the vertical axis if the bubble is still in the center. It needs to be adjusted if the bubble move off center more than one grid.

3. Adjusting steps

The bubble is brought half-way back to the center by rotating foot screws. Then the bubble should be centralized by rotating its capstan adjusting nuts. Repeat the procedures until the bubble offset is always within one grid no matter where the alidade sight to.

8.4.2 Testing and adjusting of collimation axis perpendicular to horizontal axis

1. Testing requirements

(1) The testing should be carried out by each person independently, either by horizontal ruler reading method or by quarter method.

(2) Record the readings on the ruler after direct and reverse observing, and calculate the reading difference.

(3) If the difference between the two readings is less than 3 mm, the average value can be taken as the angle distance between the collimation axis and the horizontal axis of the instrument. Otherwise, it should be tested again.

2. Testing and adjusting steps

(1) Method 1

As shown in Figure 8-1, set up an instrument and level it precisely, aim at a target A whose height is same as the height of the instrument, get the reading α_{left} on the horizontal circle when the instrument is in face left position. Then change it into face right, aim at the target A again, and get the reading α_{right} on the horizontal circle. If $\alpha_{right} = \alpha_{left} \pm 180°$, the line of sight is perpendicular to the horizontal axis. Otherwise, it needs to be adjusted.

When adjusting, the due reading of α_{right} should be calculated first, that is

$$\alpha'_{right} = \frac{1}{2}[\alpha_{right} + (\alpha_{left} \pm 180°)]$$

Rotate the tangent screw until the reading of the horizontal dial is α'_{right}, then adjust the pair of correction screws of the crosshair to aim at target A, make the longitudinal wire of the crosshair on the same vertical line as target A. Repeat the calibration until the difference ($2c$) between the face left and face right readings (the right reading should add or subtract 180°, then compare with the left reading) of the horizontal dial is less than 60″. Finally, fix the crosshair cover.

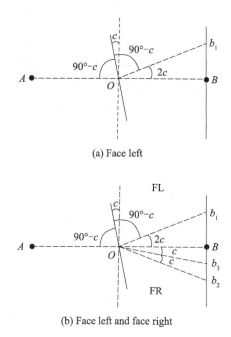

(a) Face left

(b) Face left and face right

Figure 8-1 Principle of the collimation axis testing and adjustment

(2) Method 2

As shown in Figure 8-1, select two points A and B with a distance about 100 m to each other on the ground, set up a theodolite at the midpoint O between A and B, put a target (such as a prism) at the same height as the theodolite at point A, put a horizontal ruler at point B at the same height as the instrument. The ruler is perpendicular to OB direction in the plane. Aim at the target on point A in face left position (horizontal clamp screw should be tighten), transit the telescope and get mark b_1 on the horizontal ruler. Then turn the instrument to face right, repeat the procedure, and get mark b_2 on the horizontal ruler. If $b_2 = b_1$, it means that the collimation axis is perpendicular to the horizontal axis, otherwise it needs to be corrected.

Turn a pair of correction screws of the crosshair so that the reading on the longitudinal wire comes to b_3. Repeat the above steps until the requirement is met.

$$b_3 = b_1 + \frac{3}{4}(b_2 - b_1)$$

Adjusting requirement: it should be corrected if the average value of difference between the two readings is greater than or equal to 8 mm.

8.4.3 Testing and adjusting of crosshair

1. Crosshair testing

(1) Testing requirements

① Everyone should finish this test independently, and record the deviation length (estimated in "mm") of the vertical wire from one end to the other of a fixed point.

② Describe the state of the instrument with the approximate results of two tests.

(2) Testing steps

Aim at a sharp point with the intersection of the crosshair, rotate the tangent screw of telescope to move the vertical hair up or down. If the point never leaves the vertical hair, it means that the vertical hair is perpendicular to the horizontal axis, otherwise it needs to be corrected.

2. Crosshair adjusting

(1) Adjusting requirements

① If one end of the vertical hair deviates from the fixed point by more than 2 mm according to the two testing results, it should be corrected (Figure 8-2a).

② After adjustment, the vertical hair should not deviate from the fixed point.

(2) Adjusting steps

When adjusting, unscrew the diaphragm, use a small screwdriver to loosen the four fixing screws of the outer crosshair ring (Figure 8-2b), rotate the reticle ring until the vertical hair stays on point P while moving the telescope in altitude (Figure 8-2c), and finally tighten the fixing screws of the outer crosshair ring.

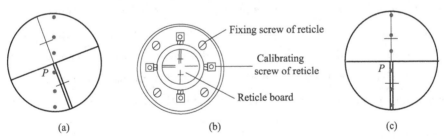

Figure 8-2 Calibration of crosshair

The horizontal hair and vertical hair are perpendicular to each other, if vertical hair is set vertical, then the horizontal hair is also horizontal.

The inspection of the collimation axis and the crosshair should be carried out separately. Adjusting of these two items can be carried out at the same time until each item has obtained a satisfying outcome.

8.4.4 Testing and adjusting of optical plummet

1. Testing requirements

The collimation axis of an optical plummet should be coincide with the vertical axis of the instrument after being refracted by the right-angle prism. If it does not coincide, alignment error will occur when the alidade is rotated. Schematic diagram of an optical plummet is shown in Figure 8-3.

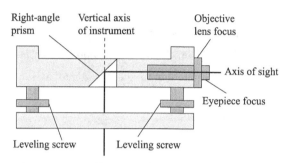

Figure 8-3 Schematic diagram of an optical plummet

Testing requirements of the optical plummet are as follows:

(1) Everyone should test it independently and make records.

(2) The testing results of the two persons determine whether the instrument meets the ideal relationships.

2. Testing and adjusting steps

First, set up and precisely level an instrument. And then fix a piece of paper on the ground below the instrument, draw a mark on the paper at the intersection of the optical alignment axis and the paper. Rotate the alidade horizontally through 180° in azimuth and make a second mark. If two marks coincide, the plummet is in adjustment. If not, the correct position of the plummet axis is given by a midpoint between the two marks. Consult the instrument handbook, adjust either the crosshairs or objective lens on the optical plummet.

8.5 Precaution

(1) Cherish instruments and do not move the screws of the instrument arbitrarily.

(2) Test and adjust an instrument according to the experimental steps. The testing order cannot be reversed. The correction can be carried out only when the inspection data is correct.

(3) The data of the instrument and the adjusted method should be explained to instructor if any part of the instrument needs to be adjusted, and the correction should be carried out after approval.

(4) The correction should be carried out under the guidance of instructor, and the testing

and adjusting should be repeated until all the requirements are met, each correction screw should be slightly tightened at the end of correction.

(5) Fill in the Table 8-1 carefully and submit it as the experiment result.

Table 8-1 Testing and adjusting of thedolite

Class: _____ Group: _____ Date: _____ Observer: _____

Instrument tag number: _____ Recorder: _____

Testing items	Testing procedure	
	Diagram	Data and description
Axis of the plate bubble ⊥ vertical axis		
Circular level bubble axis // Vertical axis		
Horizontal hair ⊥ Vertical axis		
Collimation axis ⊥ Horizontal axis		
Horizontal axis ⊥ Vertical axis		

Experiment 9　Steel Tape Measurement

9.1　Purpose and requirement

9.1.1　Experiment purpose

(1) Master the essentials and calculation methods of distance measurement.

(2) Master the method of line alignment with range rod.

9.1.2　Experiment requirement

The relative error of distance should be less than 1/2 000.

9.2　Arrangement and equipment

(1) Experimental arrangement: 2 class hours for this experiment, 3 people in each group.

(2) Experimental equipment(per group): 1 steel tape (30 m), 3 range rods, 4 chaining pins and 1 recording board.

9.3　Methodology and procedure

Measuring distance by steel tape is the basic method of distance measurement. Before measuring, mark the two endpoints of the distance to be measured with wooden stakes (with a small nail at the top of the stake). Usually, after removing obstacles on this straight line, two people set the line and measure between this two points, and the specific operation method is as follows:

(1) As shown in Figure 9-1, when measuring a distance, first set range rods (or chaining pins) at points A and B, and assign the direction of the straight line. The zero end of the steel tape held by the rear tapeperson is located at point A. The front tapeperson holds the end of a steel tape and carries a bundle of chaining pins, heads along the AB direction and stops at a length of one tape length. Both of the tapepersons squat down.

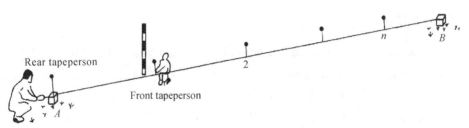

Figure 9-1　Distance measurement on flat ground

(2) The rear tapeperson uses a gesture to command the front tapeperson to pull the steel tape in the straight direction of AB. After the two persons have tightened, flattened and stabilized the steel tape at the same time, the front tapeperson shouts "preparation", and the rear tapeperson points the steel tape's zero point precisely to A and shouts "OK". Then the front tapeperson inserts the chaining pin to get point 1 (if the ground is too hard to be inserted in, pencil can be used to draw marks on the ground). The measurement of the first section is hence finished.

(3) The rear tapeperson and the front tapeperson hold up the steel tape and move forward together. When the rear tapeperson reaches point 1, he shouts "stop", and they two measure the second section in the same way. Then the rear tapeperson pulls out the chaining pin inserted in point 1 and take it with himself. In this way, the distance between A and B can be measured part by part until its last part is shorter than the full length of the steel tape. If the length of the last part is marked by q, then the horizontal distance between point A and B will be

$$D = nl + q \tag{9-1}$$

In the formula: n—the number of full length segments

(i.e., the times of chaining pins pulled by the ruler between points A and B);

l—tape length, m;

q—partial tape length, m.

In order to prevent measurement errors and improve accuracy, a return measurement should be carried out from B to A, and a new alignment should be assigned.

$$D_{av} = \frac{1}{2}(D_f + D_b) \tag{9-2}$$

In the formula: D_{av}—average distance measured back and forth, m;

D_f—horizontal distance surveyed from A to B, m;

D_b—horizontal distance surveyed from B to A, m.

The accuracy of a distance is usually expressed by the relative error K, which is converted into a fractional form with a numerator of 1, that is

$$K = \frac{|D_f - D_b|}{D_{av}} = \frac{1}{D_{av}/|D_f - D_b|} \tag{9-3}$$

9.4 Precaution

(1) In this experiment, realignment is required for surveying distance back. The chaining pins should be inserted straight, and the times of chaining pins inserted should be recorded correctly. The alignment of each section should be accurate, so that the distance can be measured in a straight line.

(2) Pay attention to the position of the zero mark and end mark of a steel tape, the notes of

meters and decimeters when measuring a distance can help prevent misreading.

(3) A steel tape must be straight and the two ends should be held at the same elevation, maintaining a steady pull. In case of slopes or potholes, the horizontal distance should be measured directly by using range poles or hanging balls. In this case, special attention should be paid to projecting the end of the steel tape on the ground.

(4) In order to keep the steel tape intact, do not kingking, fold, press or drag the tape. After use, it should be wiped before being twisted into circular loops.

(5) Fill in the Table 9-1 and submit it as the experiment result.

Table 9-1　Record of measured distance by steel tape

Class: _____　　Group: _____　　Date: _____　　Observer: _____
Instrument tag number: _____　　Recorder: _____

Section	Observation times	Number of full tape length	Partial tape length/m	Distance/m	Average distance/m	Relative accuracy K	Temperature/℃
	D_f						
	D_b						
	D_f						
	D_b						
	D_f						
	D_b						
	D_f						
	D_b						
	D_f						
	D_b						
	D_f						
	D_b						
	D_f						
	D_b						
	D_f						
	D_b						
	D_f						
	D_b						

Part Ⅰ Civil Engineering Surveying Experiment

Experiment 10 Use of Total Stations

Use of total stations

10.1 Purpose and requirement

(1) Study the functions of a total station and the names of its main components.

(2) Master the basic operation method of total station, and practice observations of horizontal angle, vertical angle and distance.

(3) Be familiar with coordinate measurement of digital mapping and points staking out during construction.

10.2 Arrangement and equipment

(1) Experimental arrangement: 4 class hours for this experiment. 2~3 people in each group. Group members operate the instruments and record data in turn.

(2) Experimental equipment(per group): 1 total station of 2″, 1 prism or 5~6 reflectors, 1 steel tape of 2 m.

10.3 Methodology and Procedure

10.3.1 Preparations

(1) First, understand the structure of total station. The structure of total station is shown in Figure 10-1.

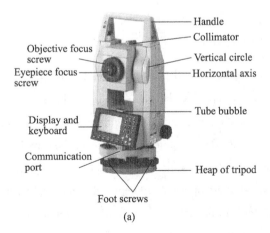

(a)

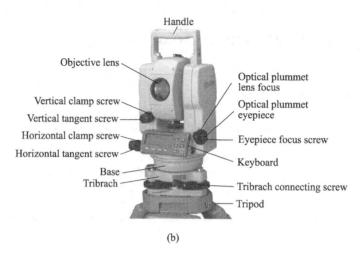

(b)

Figure 10-1 Structure of total station

(2) Install a battery, press "ON" to start a machine self-checking. Loosen the horizontal and vertical brake screws, rotate the alidade in the horizontal direction and rotate the telescope in a vertical direction for one round, then the indicators of the horizontal dial and the vertical dial are set, and "measurement mode" screen appears.

"F1—F4" in the bottom of the window are function keys (also known as "soft keys"), which are required to realize various functions of a total station.

It is necessary to input some numbers or letters according to the indication of cursor in the display window when applying function of a total station. In addition, the "Backspace" key can delete entered characters on the left of the cursor; "Esc" key can cancel the entered information.

10.3.2 Set up total station

A tripod is placed over an occupied point, the head of the tripod is almost horizontal, and the instrument is placed on the top of it, hold the handle of the instrument with one hand and fix the connecting screw with the other hand. Move the tripod, make the center of the optical plummet and the occupied point on the same plumb line, adjust the length of the tripod legs and center the circular bubble. Rotate the alidade so that the axis of the tubular bubble is parallel to a pair of foot screws, and use the foot screws to center the bubble. Turn the alidade by 90°, and center the bubble with the third foot screw. Check whether the instrument is accurately centered. If necessary, slightly loosen the connecting screw, move the instrument to make accurate alignment, and then check whether the bubble of the level tube is still in the center. Repeat the above steps until the instrument is centered and leveled.

10.3.3 Angle measurement

1. Horizontal angle measurement

As shown in Figure 10-2, usually sight to a left target L from station S in the direct position (face left), and press the offset button in the measurement mode to make the original reading of the horizontal dial 0°00′00″. Turn the alidade and sight to a right target R, and the reading of the horizontal dial is assumed 106°16′20″, which is the horizontal angle α. The horizontal angle is measured in the same way in the reverse position (first rotate the alidade 180°, and then rotate the telescope 180° around the horizontal axis, the instrument is changed into face right position). If the difference between the two measured horizontal angles is not greater than 42″, the average value is considered as one round of the horizontal angle α. If the difference is greater than 42″, the angle needs to be remeasured.

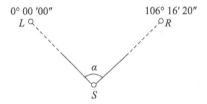

Figure 10-2 Horizontal angle measurement

2. Verticle angle measurement

If the vertical angle of a target needs to be measured at the same time, it can be read from the window when aiming at the target in the angle measurement mode. As shown in Figure 10-3, when aiming at target L the reading is 49°59′50″, the vertical angle β is 90°−49°59′50″= 40°00′10″(The vertical disk in marked clockwise).

The vertical angle is measured in the same way in the reverse position (face right position). Now the vertical angle $\beta = R - 270°$, and R is the vertical reading in face right position. If the difference between the two measured vertical angles is not greater than 42″, then the average value is considered as the vertical angle β in one round. If the difference is greater than 42″, the angle needs to be remeasured.

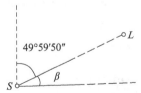

Figure 10-3 Vertical angle measurement

10.3.4 Distance measurement

1. Parameter setting

Before distance measurement, the following settings should be completed: measurement mode, target type, prism constant correction, temperature and atmos. The names and meanings of parameters are as follows:

(1) Mode with options of Fine "r" (repeated precision measurement) / Fine AVG (average precision measurement) / Fine "S" (single precision measurement) / Rapid "r" (repeated rough measurement) / Rapid "s" (single rough measurement) / Tracking (tracking distance measurement).

(2) Reflector (target type), with options of Prism or Sheet.

(3) PC (prism constant), whose value is input according to the prism used (unit: mm), refer to manuals of the instruments.

(4) Temp (temperature), input the value according to the temperature at the time of measuring (unit: ℃).

(5) Press(pressure): input the value according to atmos at the time of measuring (unit: hPa).

(6) PPM (meteorological correction percentage), calculated automatically by the instrument.

2. Distance measurement procedures

Aim at a target with a prism or a reflector on it in the measurement mode, press "Distance measurement" after aiming at the target, press "measure" on the soft keyboard, the instrument starts to measure the distance according to the set parameters. After surveying, the screen displays the values of slope distance(SD), vertical distance (VD) and horizontal distance(HD) immediately.

If the ranging mode is set to "single precision measurement", the measurement will stop automatically after the first time is completed. If it is set to "average precision measurement", the values of each time will be displayed as S-1, S-2, ⋯, S-9. After the measurement is finally completed, the average value of the distance will be displayed. The latest measured values of distance and angle are automatically saved in memory and can be gotten at any time, but the saved data will be cleared after power off.

10.3.5 Coordinate measurement

A three-dimensional(3D) coordinates (X, Y, Z) measured by total station is used for data acquisition of topographic survey. According to known coordinates (or azimuth) of a station and a BS point, the three-dimensional coordinates of the observation point are measured through distance and angle observation, and stored in the memory.

The steps are as follows:

1. Select working file

Select or create a file as the "current working file" to record results. Move the cursor to the file name to be selected (or input the new file name) and press "Enter", then the file can be selected as the data collecting file of the current project (current working file).

2. Input the data of the station

As shown in Figure 10-4, O represents the occupied point, A represents the backsight point, P represents the foresight point, and OPQ represents the forward direction of the observation. In coordinate measurement mode, input the following data: coordinates of the occupied point (N_0, E_0, Z_0), instrument height (Inst. h.), target height (Tgt. h.) and coordinates of the backsight (X_A, Y_A), press "Enter" after inputting each piece of the data. This information is recorded in the selected file.

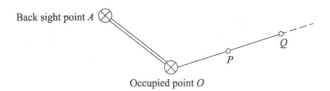

Figure 10-4 Schematic diagram of coordinate measurement

3. Measure 3D coordinate

After inputting the relevant data of a station, aim at a backsight point and press "Enter". The value displayed thereafter is the azimuth from the survey station to the backsight point calculated by the total station automatically. Return to the "coordinate measurement menu", aiming at the target P, the three-dimensional coordinates (N_P, E_P, Z_P) of point P can be determined through the measurement of slope distance S, zenith ZA and target azimuth HAR. The calculation formula are as follows:

$$\begin{cases} N_P = N_0 + S\sin ZA\cos HAR \\ E_P = E_0 + S\sin ZA\sin HAR \\ Z_P = Z_0 + S\sin ZA + h_I - h_T \end{cases}$$

In the formula, h_I is the instrument height, h_T is the target height. The calculation is completed by the instrument automatically and displayed on the screen, and can be recorded in the selected working file.

After aiming at the prism of the target, press the button "Measure" in the coordinate measurement mode, the three-dimensional coordinates, distances and angles of the target point are displayed immediately. If the measurement is continued, move the instrument to point P in Figure 10-4, where point P is the occupied point, point O is the backsight point. Repeat the above steps to determine the coordinates of the next point.

10.3.6 Setting-out

Setting-out is to mark points specified by known design coordinates (or direction and distance) in the field. The instrument can display the difference between the preinput value to be set out and the measured value by measuring the horizontal angle, distance or coordinate of the point where the prism is located during setting-out, the formula is

$$\text{Display difference} = \text{Measured value} - \text{Setting-out value}$$

According to the difference, the prism on the target point moves in order to make the difference less than the allowable misclosure, and finally find the point specified by the design coordinates accurately. Setting-out is usually carried out at face left position.

1. Angle and distance setting-out

On a station, the designated point can be set out according to the horizontal angle rotating from a reference direction and distance.

Sight to a backsight point after placing an instrument over the station. Press the button "S. O.", set the reference direction to zero. When setting out angle and distance, the following two data should be input: distance (S. O. dist) and angle(S. O. H. ang), press "Enter" after each datum is input. Move the prism or target and keep tracking and measuring. The distance to the destination S-O-S will be displayed after observation, and the horizontal angle between the connecting line from the position of the prism to the station and the reference direction dHA will be displayed. During the measurement, as the prism is moved, the setting-out data on the display will change and gradually approach the precise value until the designated point is found.

2. Coordinate setting-out

Coordinate setting-out is to mark points whose coordinates are given in the designated field. Assume that the coordinates of a station and a backsight point are known, after inputting the coordinates of the points to be set out, the instrument can automatically calculates the required azimuth and offset and stores them in memory. Using the functions of angle and distance setting-out, the position to be set out can be fixed.

The "setting-out menu" is displayed after pressing the button "S. O." in the measurement mode. Input the three-dimensional coordinates (N_0, E_0, Z_0) of a station, the instrument height (Inst. h.) and the target height (Tgt. h.). Press the button "Enter" after inputting each datum, press "OK" after inputting all data, return to the "setting-out menu", the setting of this station is then completed. Input the three-dimensional coordinates of the point to be set out, the screen will display the distance difference and azimuth difference between the target point and the current position, as shown in the Figure 10-5.

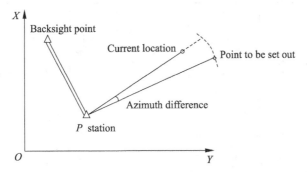

Figure 10-5 Coordinate setting-out

During observation, the setting-out data displayed on the screen also changes with the movement of the prism, gradually approaching to the destination. Finally, use tangent screw to adjust it until the difference of the distance and the angle displayed on screen are "0". Now, the exact position is observed and a mark at the corresponding position should be made.

10.4 Precaution

(1) Cherish instruments and handle them with care, protect instruments from moisture.

(2) Each person should at least complete all measurements for one station.

(3) The observation data of this experiment should be recorded in the Table 10-1 and submitted as the experimental results.

Class: _____ Group: _____ Date: _____ Observer: _____ Recorder: _____ Instrument tag number: _____
Prism height: _____ m

Table 10-1 Table for total station

No.	Mark	Instrument height/m	Traverse angle/(° ′ ″)	Distance/m	Coordinate			Coordinate increment adjustment			Adjusted coordinate		
					x/m	y/m	z/m	$\Delta x/m$	$\Delta y/m$	$\Delta z/m$	x/m	y/m	z/m
1	BS							—	—	—	—	—	—
	Station			—									
	FS												
2	BS							—	—	—	—	—	—
	Station			—									
	FS												
3	BS							—	—	—	—	—	—
	Station			—									
	FS												
4	BS							—	—	—	—	—	—
	Station			—									
	FS												
5	BS							—	—	—	—	—	—
	Station			—									
	FS												
Check													

Note: "——" refers to no data should be filled in the table.

Part Ⅰ Civil Engineering Surveying Experiment

Experiment 11 Understanding and Using of GPS

11.1 Purpose and requirement

Use of GPS

(1) Master basic operations of GPS and selection of control points.

(2) Master methods of GPS measurement and recording.

11.2 Arrangement and equipment

(1) Experimental arrangement: 2 hours for this experiment, 2 persons in each group, one operates an instrument and the other records data, and then switch roles.

(2) Experimental equipment(per group): 1 GPS (This experiment takes Situoli S5 Ⅱ as an example) equipped with 1 manual and 1 recording board.

11.3 Methodology and Procedure

Under normal conditions, the general steps for using GPS are as follows:

(1) Set up the reference station. Note: It is not necessary to set up a reference station when connecting CORS differential signal.

(2) Open SurPAD4.0 software in the manual. After the manual is connected to the reference station, create a new project, set coordinate system parameters and reference station parameters, so that the reference station can transmit differential signals.

(3) Connect and set up mobile station so that the mobile station can receive the differential data of the reference station and reach a fixed solution.

(4) The mobile station should measure the original WGS-84 coordinates of known points in the fixed solution state at a known points in a survey area. The conversion parameters between the two coordinate systems can be calculated according to the original coordinates and local coordinates of the known points, and the conversion parameters should be applied.

(5) Go to another known point to check whether the local coordinates are correct after conversion.

(6) Start survey (Base station translation and calibration have to be done for mobile station if the reference station is restarted or moved, and marker points will be used for calibration.)

(7) Export data in required format.

11.3.1 Experimental preparation

Install a SIM card (SIM cards of China Mobile, China Unicom or China Telecom are all suitable, the SIM card can access to the Internet) and memory card into the built-in slot of a

GPS instrument, install a battery, turn on the receiver and start receiving satellite signals. Using this mode, the reference station can send data to the mobile station through the mobile network.

11.3.2 Connect an instrument and set the working mode

After an instrument is powered on, perform the following operations: "Instrument"—"Communication setting", select the instrument type "RTK", select the "Bluetooth" as the communication mode, click "Search", find the name of your instrument in the bluetooth device list, and click "Connect", a connection progress bar pops out. The manual and instrument are successfully connected after the progress is completed.

11.3.3 Host self-checking

The self-checking function is mainly used to judge whether each module of the receiver works normally. The self-checking function can be used to test the receiver, when the pilot lamp of S5 Ⅱ does not light up or some modules cannot work normally. Self-checking of S5 Ⅱ includes five parts: Global Novigation Satellite System (GNSS), transmitter-receiver, Wi-Fi, bluetooth and sensor. During the self-checking, there will be voice broadcast of the testing results.

Execute "Instrument"—"Mobile Station Mode", select "Manual Network" as the data link mode, set an IP and port of the CORS server, obtain and select the access point, and use the default value for other options. Click "Apply", and the working mode is set. Return to the main page and hold the instrument stable to check whether the fixed solution is obtained.

11.3.4 Create a new project

Execute SurPad4.0, click "Project"—"Project Management"—"New", create a new project, enter a project name in a popup dialog, select coordinate parameter type, other additional information can be left blank, then click "OK", turn to the page of coordinate system parameter. Then click "Projection Parameter", select the projection method in the combo box, generally "Gaussian projection" is selected. The parameter of the central meridian can be input directly or it can be calculated and added automatically with the software in data collector (this refers to the case where the manual is connected to the GPS and the GPS has locked the satellite), as shown in the Figure 11-1.

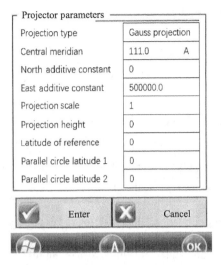

Figure 11-1 Projector parameters

11.3.5 Correction

1. Base station correction

After getting the fixed solution, execute "Project"—"Station calibration"—"Using datum point for calibration", enter the known coordinates, set the antenna parameters of the current coordinate for the base station, and click "calculate" to get the calibration parameters.

2. Point correction (solving transformation parameters)

It is necessary to make point correction if you want to match the coordinates of the known points with the point to be measured when you go to a survey area for the first time.

After an instrument is set, the parameters can be calculated or calibrated after the mobile station reaches the fixed solution. Since the original data measured by RTK is WGS-84 coordinates, the parameters must be calculated by seeking the control point of the coordinate system to be measured, and the data measured by the instrument can be converted into the coordinates under the coordinate system we need, such as National 80 or Beijing 54 coordinates system.

Detailed steps:

Set up a mobile station over a control point, hold the rod of GPS instrument and center the bubble, and collect the original WGS-84 coordinates of the control points (about 4 control points) by the function of "Point measurement". Then, execute "Tools"—"Conversion Parameters", input the coordinates of known points (select from the coordinate point file or input the value) and original coordinates in WGS84 ellipsoidal system (obtain the current GPS data or select from the coordinate file or input), decide whether to use plane correction and elevation correction, and click "OK" to complete the input of conversion parameters. Click

"Calculate" in the conversion parameter screen to get the GPS parameter report.

11.3.6 Measure

Set up the instrument over a control point, set the antenna height, after centering and leveling the instrument, execute "Measure"—"Point/Detail survey". Take "Topographical point" as an example, click "Point type", select the topographical point, and then click "setting" to set the limit conditions (example: fixed solution, H:0.05, V:0.1, PDOP:3.0, postpone:5, smooth:1) of topographic point. Click the button "Data receiver" on the bottom right to complete the acquisition and save the data.

11.3.7 Export data

Export data from the manual and copy them to a computer, the data can be processed in office.

11.4 Precaution

(1) There should be no shelter or electromagnetic interference (no microwave stations or radar stations or mobile phone signal stations, etc. within 200 m, and no high-voltage lines within 50 m) near the base station.

(2) There are at least two known coordinate points (the known points can be in any coordinate system. It will be better if there are three or more points so that they can be checked with each other).

(3) The observation data of this experiment should be recorded in the Table 11-1 and submitted as the experimental results.

Table 11-1　Data log sheet for GPS

Class: _____　Group: _____　Date: _____　Observer: _____

Instrument tag number: _____　Recorder: _____

Latitude: _____　Longitude: _____　Height: _____　Temperature: _____ ℃
Pressure: _____　Relative humidity: _____　PDOP: _____　Initial slant height: _____ m
Final slant height: _____ m
Description of historical site (include rubbing of surface):
Potential problems:
Contact situation to historical site:
Comments:

Station	Coordinate		
	X	Y	Z

Experiment 12 Staking out Axis and Elevation of Building

12.1 Purpose and requirement

(1) Master the basic methods of staking out building axis.

(2) Master the basic methods of staking out elevation during construction.

12.2 Plan and equipment

(1) Experimental arrangement: 2 class hours for this experiment, 3 persons in each group.

(2) Experimental equipment(per group): 1 total station, 1 level, 1 steel tape, 1 range rod, 1 leveling rod, 1 recording board, 1 hammer, 6 wood piles, 2 surveying pins. Calculator is self-provided.

12.3 Methodology and procedure

12.3.1 Control points layout and data calculation

Control points are necessary for staking out building axis (see Figure 12-1) and elevation. For this purpose, select two points A and B on the open ground, first drive a wooden pile as point A, draw a cross line on the top of the pile, take the intersection as the center, and measure a distance of 50.000 m with a steel tape to determine point B (also drive a wooden pile, draw a cross line on the top of it). The coordinates of points A and B are assumed to be A ($x_A = 100.000$ m, $y_A = 100.000$ m); $B(x_B = 100.000$ m, $y_B = 150.000$ m).

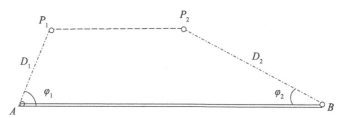

Figure 12-1 Staking out building axis

Suppose the elevation of point A is 10.000 m and the data above are known for existed control points. Coordinates and elevations of points P_1 and P_2 on the axis of proposed buildings are as follows: $x_{P_1} = 108.360$ m, $y_{P_1} = 105.240$ m, $H_{P_1} = 10.150$ m; $x_{P_2} = 108.360$ m, $y_{P_2} = 125.240$ m, $H_{P_2} = 10.150$ m. In an actual experiment, relevant data are drawn up according to the requirements of an instructor. In a practical project, staking out should be done according to the data on the design drawings.

At control points A and B, stake out the plane positions of points P_1 and P_2 (these two points are on a certain axis of the building) by polar coordinate, and set out the elevations by level.

When using the polar coordinate method to stake out point P_1, we have

horizontal distance: $D_1 = \sqrt{(x_{P_1} - x_A)^2 + (y_{P_1} - y_A)^2}$

horizontal angle with the known direction: $\varphi_1 = \arctan \dfrac{y_B - y_A}{x_B - x_A} - \arctan \dfrac{y_{P_1} - y_A}{x_{P_1} - x_A}$

The data D_2 and φ_2 can be calculated by the same method.

Calculate the required data in the Table 12-1, and draw a sketch of the building axis location.

12.3.2 Staking out axis by polar coordinate

(1) Set up a total station at point A, make it sight to point B, set the initial reading of the horizontal dial to $0°00'00''$, rotate the alidade anticlockwise so that the reading of the horizontal dial is $(360° - \varphi_1)$, mark this direction on the ground with a chaining pin, measure the horizontal distance D_1 from point A in this direction, and drive a wooden pile. Then check the direction with the total station and measure the distance, and set point P_1 on the wooden pile.

(2) Set up the total station at point B, stake out P_2 in the similar method (the difference is that the alidade is rotated clockwise for the angle φ_2 after sighting to point A)

(3) According to the horizontal distance calculated by the design coordinates of these two points, the positions of wooden piles at P_1 and P_2 can be determined. Check them with the total station, and the difference between the theoretical value and surveyed value should not be greater than 10 mm.

12.3.3 Setting out axis by rectangular coordinate method

Refer to the coordinate setting out by total station in Experiment 10 for setting-out method.

12.3.4 Staking out elevations of axis points

Set up a level in the halfway among the point A, point P_1 and point P_2, the distance from station to any points is roughly equal. Hold a level rod on the wooden pile of point A whose height is given, get the backsight reading a, according to the height H_A of point A, the height of sight (instrument height) H_i is obtained (Figure 12-2).

$$H_i = H_A + a$$

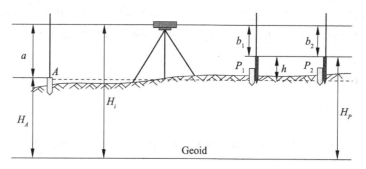

Figure 12-2 Staking out the elevation of axis point

The design height of point P is H_p, the height difference between points P_1, P_2 and point A is

$$h = H_p - H_A$$

the reading on the level rod of points P_1 and P_2 should be

$$b = H_i - H_p$$

Drive wooden piles at point P_1 and point P_2 respectively, drive them into the soil deeper and deeper gradually until the reading on the level rod equals b. The elevation of pile top is the design elevation of the axis.

12.4 Precaution

(1) Group members calculate staking out data independently, check the calculation results with each other, and stake out the positions of the points if the data are correct.

(2) The observation data of this experiment should be recorded in the Table 12-1 and submitted as the experimental results.

Table 12-1 Computation for staking out

Class: _____ Group: _____ Date: _____ Observer: _____

Instrument tag number: _____ Recorder: _____

Data record of plane survey setting-out					
Station	Coordinate or coordinate increment		Horizontal distance/m	Coordinate azimuth/(° ′ ″)	Horizontal angle/(° ′ ″)
	X (or ΔX)	Y (or ΔY)			

Data record of elevation survey setting-out					
Given elevation/m	Elevation to be set out/m	BS reading/m	Sight line elevation/m	Calculated reading of point to be set out/m	Remark

Schematic diagram of staking out

Experiment 13 Staking out Circular Curve of Road

13.1 Purpose and requirement

(1) Master the calculations of main points of circular curve and the method of staking out main points.

(2) Master the method of staking out circular curve using deflection angle.

13.2 Arrangement and equipment

(1) Experimental arrangement: 3 class hours for this experiment, 3 persons in each group.

(2) Experimental equipment (per group): 1 total station, 1 prism, 10 chaining pins, 3 wooden piles, 1 hammer and 1 recording board.

13.3 Methodology and procedure

13.3.1 Staking out the main points of curve

Before staking out the main points of a road curve, the intersection and turning point of road which mark road direction should be staking out first. As shown in Figure 13-1, make a wooden pile at the point of intersection JD_1 and extends more than 30 m in both directions, then set two turning points ZD_1 and ZD_2, insert the chaining pins at the turning points.

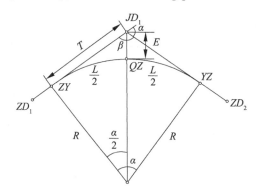

ZY—Point of curve; QZ—curve midpoint; YZ—Point of tangent; R—curve radius; L—curve length; E—external distance; T—tangent distance; β—traverse angle; α—intersection angle.

Figure 13-1 Circular curve elements

Set up a total station at the point of intersection JD_1, staking out the traverse angle β, calculate the deflection angle α ($\alpha = 180° - \beta$). Calculate the circular curve elements, according to the following formula (tangent length T, curve length L, the external distance E, difference of the tangent and curve J), and record it in the Table 13-1.

$$T = R\tan\frac{\alpha}{2} \tag{13-1}$$

$$L = R\alpha\frac{\pi}{180°} \tag{13-2}$$

$$E = \frac{R}{\cos\dfrac{\alpha}{2}} - R \tag{13-3}$$

$$J = 2T - L \tag{13-4}$$

Set up a total station over the point JD_1, sight to points ZD_1 and ZD_2 successively, then the tangent direction is fixed. Along this direction, set out the tangent length T, staking out wooden piles on the points ZY and YZ, repeat the procedure, make sure the locations of points ZY and YZ are accurate.

Make the total station sight to point YZ, set the original horizontal circle reading at $0°00'00''$, rotate the collimator for a horizontal angle $\beta/2$, along the direction of angle bisector, measure the external distance E by steel tape, and fix the midpoint QZ of this curve.

13.3.2 Calculation of stationing

Station values of main points of the curve located on the center line of the road is calculated from the station of the intersection.

$$\begin{cases} ZY \text{ station} = JD \text{ station} - T \\ QZ \text{ station} = ZY \text{ station} + L/2 \\ YZ \text{ station} = QZ \text{ station} + L/2 \end{cases} \tag{13-5}$$

To avoid mistake of the calculation, the following equation can be used to check the calculation. It means the calculation is correct if the two results are equal.

$$YZ \text{ station} = JD \text{ station} + T - J \tag{13-6}$$

13.3.3 Staking out curve in detail by deflection angle method

Assume that mileage stakes are measured every 10 m on the curve, then $l_0 = 10$ m, l_1 is arc length between P_1 (the first 10 m mileage on the curve) and ZY (the starting point of the curve), as shown in Figure 13-2.

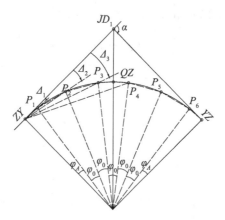

Figure 13-2 Staking out curve by deflection angle method

To stake out curve by deflection angle method, one should calculate the deflection angle Δ_1 at point P_1 and each deflection angle Δ_0 at each point which is set every 10 m arc length in the remainder after P_1 by the following formulas.

$$\Delta_1 = \frac{l_1}{2R} \times \frac{180°}{\pi} \tag{13-7}$$

$$\Delta_0 = \frac{l_0}{2R} \times \frac{180°}{\pi} \tag{13-8}$$

The deflection angles at detailed points P_2, P_3, $\cdots$, P_i can be calculated by the formula:

$$\Delta_2 = \Delta_1 + \Delta_0 \tag{13-9}$$

$$\Delta_3 = \Delta_1 + 2\Delta_0 \tag{13-10}$$

$$\cdots$$

$$\Delta_i = \Delta_1 + (i-1)\Delta_0 \tag{13-11}$$

Chord length C_i from the start point of the curve to any detailed point P_i on the curve can be calculated by:

$$C_i = 2R \sin \Delta_i \tag{13-12}$$

Chord length C_0 between adjacent station piles on the curve can be calculated according to the following formula:

$$C_0 = 2R \sin \Delta_0 \tag{13-13}$$

The difference between arc length L and chord length C (chord arc difference) between any two points on the curve is calculated according to the following formula:

$$L - C = \delta = \frac{l^3}{24R^2} \tag{13-14}$$

The steps of setting-out curve by deflection angle method are as follows:

(1) Set up a total station at ZY, sight the alidade to the point of intersection JD_1, and set the initial horizontal circle reading of the horizontal dial $0°00'00''$.

(2) Turn the alidade clockwise to set the horizontal circle reading as Δ_1. Then measure chord length C_1 from the instrument station along the sighting direction of the total station with a steel tape to locate first point P_1 on the curve and drive a chaining pin into the ground to mark it.

(3) Turn the alidade clockwise again to set the horizontal circle reading as Δ_2. Then measure the chord length C_0 from point P_1 and intersect with the sighting direction of the total station to obtain the position of point P_2 and use a chaining pin to mark. This procedure is continued until the point YZ is reached.

(4) If the total length of a curve is short, the chord length C_i can also be measured from the starting point of a curve according to the deflection direction indicated by the total station to obtain the position of point P_i.

(5) The deflection angle of YZ should be equal to $\alpha/2$. The measured length from the last point on the curve to YZ should be equal to the calculated chord length. If they do not match, the misclosure can not exceed ±0.1 m in radial direction (horizontal), and not exceed $\pm\dfrac{L}{1\,000}$ in tangent direction (longitudinal).

13.4 Precaution

(1) The staking out data of main points on circular curve and the measurement data of deflection angle should be calculated by two people independently, and the measurement can be carried out only after it is verified to be correct.

(2) This experiment needs a large site, and there are many instruments and tools which should be checked in time to prevent loss.

(3) Table 13-1 should be submitted as the results of this experiment.

Table 13-1 Field notes for horizontal curve staking out

Class: _____ Group: _____ Date: _____ Observer: _____

Instrument tag number: _____ Recorder: _____

Circular curve elements	Data	Circular curve elements	Data
Vertex JD_1		Tangent length T	
Intersection angle β		Curve length L	
Deflection angle α		External distance E	
Curve radius R		Tangential deviation J	

Station number	Curve length between adjacent stations l/m	Deflection angle $\Delta/$ (° ′ ″)	Chord length C/m	Chord length between adjacent stations C_0/m

Experiment 14 Unmanned Aerial Vehicle Topographic Mapping

14.1 Purpose and requirement

(1) Understand the composition of unmanned aerial vehicle(UAV) instrument.

(2) Get familiar with the operation of UAV, master the field data collection and establishment of 3D mapping model.

14.2 Arrangement and equipment

(1) Experimental arrangement: 2 class hours for this experiment, 5~6 persons in each group. This experiment is a demonstration experiment.

(2) Experimental equipment (per group): a set of UAV photogrammetry equipment (including battery, camera, remote controller, etc.).

14.3 Methodology and procedure

14.3.1 Technical preparation

Original topographic maps, images and data related to a survey area should be fully collected before UAV mapping. Understand the topography, climate, airports and important facilities of the survey area, study and analyze these situations. Special attention should be paid to that whether the takeoff point is forbidden area of flight.

14.3.2 Site investigation

(1) Survey a location, distribution of features and landforms, sight and road traffic conditions, humanities, meteorology, residential distribution, etc. of the survey area. According to the collected data of control point, find its location and ascertain its reliability and usability.

(2) Check Real Time Kinematic network in the survey area and ascertain the signal strength and usability. It's noted that photogrammetry should be done in sunny days. An open and unobstructed site is better to be chosen as takeoff point, so that better GNSS signal, longer communication distance, better network signal and wider vision can be obtained. The takeoff point and flight line should be far away from Wi-Fi network coverage area, signal tower, substation, etc., in which case communication link of UAV interference is avoided, which may lead to loss of control of an aircraft and loss of lock on the satellite of RTK.

(3) Premarking points with artificial targets. Premarking points should be clearly

distinguishable on the image and have distinct features and geographical coordinates. Geographic coordinates of premarking points can be obtained through GPS, RTK, total station and other measurement technologies.

14.3.3 Field data collection

1. Instrument preparation

Install UAV propeller and battery (ensure that power of a UAV battery and remote-control battery is sufficient enough to meet the requirements), check the SD card, check if the camera is clear and free of stains, if the propeller is installed correctly, and if the machine arm and footstock are fastened.

2. Planning of UVA mapping

It is difficult to control a UAV by staff in field work, so line planning software is necessary. For example, Dajiang's line planning software DJI GO and DJI FLY, as well as other softwares like Lichi and Smart Flight, etc. Before executing the task, the first step is to set the planning parameters, including task area and task type. Task area refers to an area to be photographed, and task type refers to type of modeling, such as 2D modeling or 3D modeling.

3. Parameter setting

Parameter setting includes flight height (about 100 m, determined according to the height of obstacles in the survey area), overlap rate (70%~80%), flight speed, route angle (angle of the main route is usually 90°), etc.

When aircraft RTK (installed with Qianxun CORS) positioning function is used, it should be noted that the aircraft RTK positioning function should be turned off and switched back to GNSS mode when the RTK module is abnormal. Turn on the RTK switch after the UVA takeoff, and the navigation system will continue to use GNSS mode.

Return position setting: when GNSS signal reaches four grids or more for the first time, the current position of the aircraft will be recorded as the return position. The return position can also be updated later, for example, take the current position of the aircraft, or the remote control as the return position. The return height should be higher than the highest building in the survey area.

Alarm voltage setting: it is recommended to increase the threshold of the low battery alarm voltage if aircraft flies against the wind or there have been too many battery cycle times (more than 100 times),

Route planning: the planning methods of different route planning software are slightly different. Here takes DJI GO App and DJI PILOT App as examples.

DJI GO App provides two methods for three-dimensional modeling route planning: five-way flight and zigzag flight. Five-way flight has 5 sets of routes, which include 1 orthophoto and 4 direction tilt routes, while zigzag flight is composed of 2 photogrammetric 2D flight lines which are arranged vertically, the default value of PTZ angle is $-60°$. The recommended sidelap is 80%, and the endlap is 80%. Tilt photography in DJI FLY App is five-way flight.

Press "Route Planning" button at a remote control, and choose a point on the map to plan the survey area after selecting the operation mode. A KML file made by 91 Satellite Map and other Apps of the survey area can also be used. Export the KML file to the appropriate storage of a computer, then import it into the remote control storage of the UAV through USB.

4. Task execution

After checking software and hardwares (UAV, propeller, remote controller, battery, SD card reader, etc.), turn on the power of the remote controll first, and then turn on the power of the UAV. A controller operates in the rear of the aircraft, hovers the aircraft at low altitude for about 1 min to help the battery fully warm up first, so as to ensure the normal operation of various sensors, observe whether the flight form is normal, execute the flight mission after confirmation, then the UAV starts to execute the flight line automatically.

Check the pictures and camera parameters during executing a task. The camera parameters need to be adjusted when there is overexposure or underexposure in the photos. The system will automatically record the interruption point when a single flight of UAV cannot complete the measurement job due to battery power or other reasons. Select the current task and click "Call" to continue when performing the job again.

5. Field work inspection

Check the photos taken by UAV to see whether the photos have blur, abnormal exposure and other unacceptable problems, then back up the photos. If there are any abnormal situations such as missed shots or blur, it is necessary to make up the shots in time. The reasons for missing shooting may be the followings: the selected shooting mode is "fixed distance shooting" or the reading and writing speed of SD card are slow or the SD card is full.

Import the collected data into a mapping software and perform aerotriangulation. Aerotriangulation refers to the process of using the spatial geometric relationship between the image and the captured object in photogrammetry to calculate the camera position and pose at the time of shotting and the sparse point cloud of the captured object. After processing, it can quickly judge whether the quality of the original data meets the project delivery requirements and whether the image needs to be added or deleted. Before 2D reconstruction and 3D reconstruction, aerotriangulation must be performed first. After completing the aerotriangulation, check the aerotriangulation report. If the aerotriangulation results do not meet the requirements, it is necessary to fly again, collect data, and then perform the aerotriangulation until the required aerotriangulation results are obtained.

6. Storage and maintenance of the UAV

Clean the UAV in time and store it in a safety protection box after using, and keep the environment dry. Maintain the UAV regularly.

14.3.4 Office data processing

The procedure of processing office data is shown in Figure 14-1. In the figure, POS data refers to the data that records the 3D coordinates (longitude, latitude and flight altitude) and flight attitude (heading angle, pitch angle and roll angle) of the UAV at the moment of shotting. Digital surface model (DSM), refers to a ground elevation model that includes the heights of surface of buildings, bridges, and trees. DOM stands for digital orthophoto map.

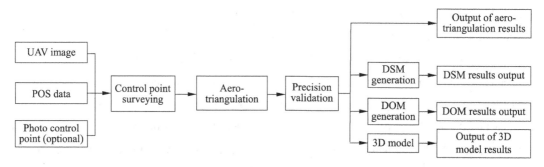

Figure 14-1 Flow chart of office data processing

1. Control point measurement

After completing the photogrammetric task, remove the SD card and transfer the data from the SD card to a computer. Create a new reconstruction task with a mapping software, import the photos taken by UAV, import camera parameters and premarking points. The premarking points are divided into control points and checkpoints. Control points are used to optimize the accuracy of aerotriangulation data, which can improve model accuracy and also achieve conversion of local coordinate systems or 85 elevation systems. Checkpoints are used to check the accuracy of aerotriangulation, and also can be used to evaluate the accuracy quantitatively.

2. Aerotriangulation measurement

First, associate the premarking points with the photos taken at those points through software pricking, then perform the calculation of aerotriangulation through software.

3. Accuracy verification

Set reconstruction parameters, such as selecting reconstruction resolution ratio and scene, etc. First, select the coordinate system and start reconstruction, after the reconstruction is completed, the reconstruction results can be displayed in the interface of map, a 2D/3D topographic map can be obtained. Then export a report, the quality of reconstruction can be understood visually by the report.

4. Results output

The orthophoto and realistic 3D model generated by DJI Terra can be imported into CASS-3D for DLG (digital line graph) production. DLG is a vector data set of map elements that are basically consistent with existing lines, it can store the spatial relationships between each element and related attribute information.

The steps for mapping based on DOM+DSM are as follows:

(1) Data preparation: Find the DOM (result. tif) and DSM (dsm. tif) results and their supporting files with filename suffixes ". prj" and ". tfw" in the engineering folder of the 2D reconstruction of DJI Terra.

(2) Model conversion: Open the CASS-3D software, click the "Build Dsm" button, enter the DOM, DSM, and storage path in the popup dialog and select the appropriate clarity (the larger the value is, the higher the clarity is, and the longer the time it takes). Click the "Start" button to generate an ". osgb" file under the set path.

(3) Open the model: Click the "3D" button to open the ". osgb" file converted in the previous step. This file is a vertical model that combines DOM with DSM, and can be used for DLG making and other operations based on this model.

14.4 Precaution

(1) The lens can be preheated through turning on the machine to accelerate water vapor dissipation if there is fog inside.

(2) Always pay attention to the change of battery.

(3) Keep the flight attitude stable and pay attention to the changes of flight environment (avoid flying in areas with strong snow reflection, extreme cold weather, snowstorm weather, etc.).

(4) During surveying, the team staffs should establish reliable communication with the local air traffic control bureau and report the flight plan. Flight plan reporting generally includes: flight preparation and estimated work time of UVA surveying; the stop time (which should be notified to the bureau at the end of the flight on that day). The detailed information of the time and the content should be implemented according to the requirements of the approval reply.

Part Ⅱ Civil Engineering Surveying Practice

 Purpose of surveying practice

Civil engineering surveying practice is an important practical part of civil engineering and related disciplines. The purpose of practice is to consolidate the students' understanding and mastery of the theoretical knowledge learned in class, master the basic operating skills of measuring instruments, focus on enhancing the students' performing and problem-solving abilities, as well as mastering the basic methods of large-scale topographic mapping.

 Practice content and requirement

Mapping of large-scale topographic map (1 : 500)

Select an area for control survey as needed, and delimit a 150 m×150 m area in it used for topographic map, a 1 : 500 topographic map should be measured and drawn according to the required accuracy of the scale.

(1) Horizontal control survey

According to the principle of selecting points for field survey, select a suitable number of control points with proper location to form a loop traverse, draw a sketch, and make a note of the points. Measure and calculate the coordinates of each control point according to the required accuracy of traverse as the plane control survey in the survey area. Allowable angle misclosure $f_\beta = \pm 60'' \sqrt{n}$ (where n is the number of measuring stations), the allowable relative error of traverse $K \leqslant 1/2\,000$.

(2) Vertical control survey

In the selected survey area, according to the principle of selecting control points of field work, some appropriate benchmarks should be arranged. The points can be the same as the horizontal control survey, sketches should be drawn, points should be recorded, and a closed leveling line should be formed. Vertical control survey should be carried out according to the requirements of leveling, and the verification should be carried out by using the twice instrument height method or the double-faced level rod method. The misclosure $f_{h\,\text{allowable}}$ should meet the requirements: $f_{h\,\text{allowable}} = \pm 12\sqrt{n}$ mm (n is the number of stations); or $f_{h\,\text{allowable}} = \pm 20\sqrt{L}$ mm (L is the total length of level route, unit: km).

(3) Topographic mapping

Measure the detail points and their elevations based on the known control points; delineate a suitable drawing area on A2 drawing and draw a coordinate grid (the total size of the grid is 300 mm×300 mm, and 100 mm×100 mm for each square). The control points and detail points should be drawn on A2 drawing. The surface features of various types are strictly represented by schematic symbols. Draw contour lines with 0.5 m as the basic contour interval (the coordinates and elevation of starting point can be assumed), and complete a large-scale topographic map in the scale of 1∶500 within the limited area. A title, a scale, an elevation system, a coordinate system, legend, etc. should be marked on the drawing. The title, the scale and the person who complete the drawing should be indicated in the countersignature column.

(4) Data calculation and setting-out steps

Select one point in the topographic map, calculate the setting-out data with polar coordinate method according to the two known control points, and explain the setting-out method, or calculate the coordinates of the selected point according to the relevant information on the topographic map, complete the staking out work based on calculated coordinates combined with the two known control points by total station.

 Requirement for practice results

(1) A leveling survey record and calculation sheet (A4 paper);

(2) A traverse survey record and calculation sheet (A4 paper);

(3) A topographic map (all landmarks in the grid should be drawn), A2 drawing (with countersignature column);

(4) A calculation process and results of setting-out data of 2~3 feature point, and explaination of setting-out method (one for each person) (A4 paper);

(5) At least 6 internship logs and a summary (for each group member) should be written on A4-size papers.

It should be noted that the contents and bookbinding requirements of each group's materials are as follows: all the materials for practice should be bound after adding the cover and directory in the following order: leveling survey record and calculation tables, traverse survey record and calculation tables, data calculation process and results of setting-out points, internship log, internship summary.

 Internship schedule

The total duration of surveying practice is 2 weeks.

Internship organization

1. Instruments and tools

For each group: a theodolite or a total station; a level; 2 leveling rods; a 50 m (or 30 m) steel tape; 2~3 range rods; a can of paint; 5~8 chaining pins; 1~2 notebooks (self-prepared); calculation table.

2. Personnel organization

Students in each class is divided into several groups, and the number of persons is equally distributed as far as possible (3~5 people in each group). Each group has a leader who is responsible for the arrangement and management of the whole group's practice. Group members must act in close coordination, help each other to complete the internship on time and with high quality.

 Grading

Internship grades are recorded in the student transcript as a separate course. The scoring content includes internship performance, report, and topographic map. The details are as follows:

1. Internship performance (20%)

Basic operation skills of level and theodolite (or total station); attitudes to the internship, following the internship discipline, unity and collaboration among team members.

2. Internship report (50%)

The calculations and data submitted by each group should be true and correct, the errors should be less than allowable misclosures; the content of the logs and summaries should be truthful and carefully written, no less than 300 words in each log and no less than 2000 words in the summary.

3. Topographic map (30%)

The topographic map submitted by each group should be accurately described, clearly marked and signed by each member of the group.

 Precaution

(1) Control points and benchmarks should be intervisibility, marked with paint (use the group logo) and named the point number, and should not be arranged in the middle of a road in order to prevent from damaging.

(2) The original measurement data should be recorded accurately and clearly.

(3) Draw the grid lightly to keep the drawings neat, complete and standardized.

(4) Pay attention to personal and property safety.

土木工程测量实验与实习要求

测量实验与实习须知

（1）在实验、实习前，应预习本书相应测量实验项目内容，明确实验、实习目的与要求，熟悉实验步骤等，并准备好必要的表格和文具等。

（2）应遵守上课纪律，不得无故缺席或迟到早退。

（3）应认真完成指导老师所布置的任务。

（4）应在指定的地点进行，不得擅自改变地点。

（5）应爱护仪器工具，严格按照测量仪器的使用规范操作。

（6）应严格按照测量资料记录规则记录。

（7）应爱护花草树木和农作物，不得任意损坏。

测量资料记录规则

（1）实验记录直接填写在规定的表格中，不得先用另纸记录，再进行转抄。

（2）所有记录和计算须用 H 或 2H 铅笔书写，不得使用钢笔、圆珠笔或其他笔书写。

（3）字迹应清晰，并书写在规定的格子内，格子上部应留有适当的空白，作错误更正之用。

（4）写错的数字用横线划去，并在原字上方写出正确数字。严禁在原字上涂改或用橡皮擦拭。

（5）严禁连续更改数字，例如，改了观测数据后，又改其平均数。观测数据的尾数原则上不得更改，如角度的度、分、秒，以及水准和距离的厘米、毫米。

（6）记录的数据应齐全，数据中的 0 不得随便省略。

（7）当一人观测、另一人记录时，记录者应将所记数字回报给观测者。

测量仪器使用规则

土木工程测量实验与实习中使用的仪器多为精密、贵重仪器。为保证仪器安全，延长其使用寿命及保持仪器精度，使用仪器时，须按以下要求进行：

（1）对光学仪器要严格防潮、防尘、防震，在雨天及大风沙天气下不得使用。

（2）仪器应尽可能避免架设在交通要道上，在架好的仪器旁必须有人看守。

（3）在架设好仪器后，必须检查脚螺旋及连接螺旋是否拧紧。

（4）若要在使用过程中近距离搬动仪器，应将制动螺旋松开。若搬动经纬仪，则还要使望远镜竖直，并将仪器抱在胸前，一手扶住基座部分，不得将仪器扛在肩上。

（5）拧动仪器的各个螺旋时，用力要适度。在未松开制动螺旋时，不得转动仪器的照准部及望远镜。

（6）工作时不得坐在仪器盒上。仪器装在盒内搬运时，应该先检查搭扣是否扣好，皮带是否结实。

（7）在使用过程中，如发现仪器转动失灵，或有异样声音，应立即停止工作，对仪器进行检查，并报告实验室相关工作人员。

（8）仪器的光学部分如沾有灰尘，应用软毛刷刷净，不得用不洁及粗糙的布类擦拭，更不得用手擦拭。

（9）使用仪器后，应详细检查仪器的状况，查看配件是否齐全。

（10）仪器装箱时应将其放置在原来的位置，且将制动螺旋放松。如果仪器箱不能盖严，不能用力按压，应检查仪器的放置位置是否正确，若不正确，将其调整到位后，再把仪器箱的盖子盖好。

（11）使用钢尺时，切勿在打卷的情况下拉尺，并且避免脚踩、车压；钢尺用完后，必须先将其擦净、上油，再卷入盒内。

（12）丈量距离时，应在钢尺卷起1~2圈的情况下拉尺，且用力不得过猛，以免将连接部分拉坏。

（13）花杆及水准尺应该保持其刻划清晰，没有弯曲，不得用来扛或抬物品及乱扔乱放。

第一部分 土木工程测量实验

实验1 自动安平水准仪的使用

水准仪的操作

1.1 实验目的

(1) 了解自动安平水准仪（DSZ3 型）的结构及工作原理。

(2) 练习自动安平水准仪的安置、瞄准操作，掌握其读数和高差计算方法。

1.2 实验安排和设备

(1) 实验安排：实验课时为 1~2 学时；实验小组由 3 人组成，其中，1 人操作仪器，1 人立水准尺，1 人负责记录。

(2) 实验设备（以小组为单位）：DSZ3 型自动安平水准仪 1 台，水准尺 2 支，记录板 1 块。学生自备 2H 铅笔和计算器。

1.3 实验方法和步骤

1.3.1 自动安平水准仪的结构及工作原理

图 1-1 所示为 DSZ3 型自动安平水准仪的结构示意。

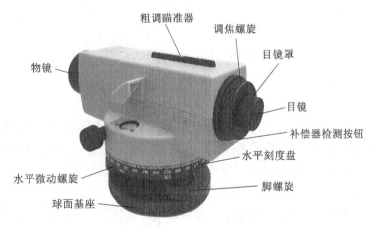

图 1-1 DSZ3 型自动安平水准仪的结构

自动安平水准仪利用圆水准器粗平仪器，仪器中的补偿棱镜在地球重力的作用下使仪器视准轴自动水平（即完成精平）。

1.3.2 水准仪的安置和水准测量

1. 安置三脚架和连接仪器

安置测量仪器的点称为测站。在选好的测站上松开脚架伸缩螺旋，按需要调整架腿的长度后将螺旋拧紧。安放三脚架，使架头的高度比肩膀的高度稍微低一点儿，并使架头尽可能水平，踩实脚架。然后把水准仪从箱中取出，放到三脚架的架头上，一只手握住仪器，另一只手将三脚架架头上的连接螺旋拧入仪器基座内固定，并用手轻轻试推仪器，检验是否连接牢固。

2. 整平水准仪

自动安平水准仪的整平是通过转动脚螺旋使圆水准器气泡居中而实现的。如图1-2a所示，当气泡未居中并位于 a 处时，可按图中所示方向用两手同时相对转动脚螺旋①和②，这时气泡移动的方向与左手大拇指的移动方向一致；若此时气泡移动到 b 处，则用左手转动另一个脚螺旋③，使气泡居中，如图1-2b所示。重复以上操作，直至转动望远镜在任何方向时气泡都居中。

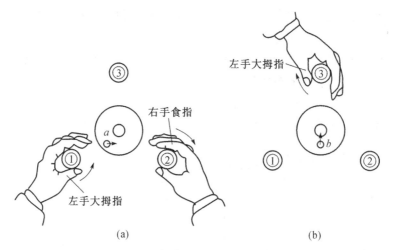

图1-2 自动安平水准仪的整平

3. 瞄准水准尺

进行水准测量时，用望远镜瞄准水准尺的步骤如下：

（1）目镜调焦。把望远镜对着明亮的背景，转动目镜调焦螺旋，使十字丝清晰。

（2）粗瞄目标。转动仪器的照准部，使目镜和准星的连线对准水准尺。

（3）物镜调焦。从望远镜中观察，转动物镜调焦螺旋，使水准尺成像清晰。

（4）精确瞄准。转动水平微动螺旋，使十字丝纵丝照准水准尺中央或边缘。

（5）消除视差。眼睛在目镜端上下微微移动，若发现十字丝横丝在水准尺上的读数也随之变化，则称这种现象为视差。这是因为尺像与十字丝分划板平面没有重合，会产生明显的读数误差，所以必须消除。消除视差的方法是仔细转动目镜调焦螺旋与物镜调焦螺旋，直至尺像与十字丝分划板平面重合。

4. 读数

瞄准水准尺后，应先整平水准仪，再读取十字丝中丝（横丝）在水准尺上所截取的数值。读数之前，一定要记得按一下补偿器（视仪器型号酌情操作）。读数时，从上向下（倒像望远镜），由小到大，并估读到 mm，读四位数。需要注意的是，不同型号的水准尺在细节上有所不同，应根据实际情况研判。

综上所述，水准仪的基本操作程序可以简单地归纳如下：安置仪器—整平—瞄准—读数。

1.3.3 水准测量记录

每人练习水准仪的安置和操作后，对两支竖立的水准尺分别进行瞄准、读数，并记录在表 1-1 中，计算两水准尺立尺点的高差。所有测站的水准测量完成后，把该记录表作为本次实验成果交给指导老师。

1.4 注意事项

（1）仪器安放到三脚架架头上后必须立即旋紧连接螺旋，确保连接牢固，以免仪器从脚架上摔下损坏。

（2）瞄准目标，在读数前必须消除视差。

（3）从水准尺上读数必须读到四位数：m，dm，cm，mm。不到 1 m 的读数，第一位数为零，如为整分米、整厘米读数，相应的位数也应补零。

表 1-1 水准测量记录

班级：_____　　组号：_____　　日期：_____　　观测者：_____

仪器编号：_____　　记录者：_____

测站	点号	水准尺读数/mm		高差/m	平均高差/m	备注
		后视	前视			

实验 2　工程水准测量

2.1　实验目的

（1）掌握水准测量中测站和转点的选择方法，并学会水准尺的立尺方法，掌握仪器的操作方法。

（2）掌握采用两次仪高法进行工程水准测量的施测、记录、高差闭合差调整和高程计算的方法。

2.2　实验安排和设备

（1）实验安排：实验课时为 2 学时；实验小组由 4 人组成，其中，1 人操作仪器，1 人记录，2 人立水准尺。

（2）实验设备（以小组为单位）：水准仪 1 台，水准尺 2 支，尺垫 2 个，记录板 1 块。学生自备 2H 铅笔和计算器。

2.3　实验方法和步骤

2.3.1　两次仪高法

两次仪高法也叫变动仪高法，是在同一个测站上用两次不同的仪器高度，将两次测得的高差进行检核。即测得第一次高差后，改变仪器的高度（改变量大于 10 cm）再测一次高差，两次所测得的结果之差不超过容许值（普通水准测量为 ±5 mm），取其平均值作为所测结果。

2.3.2　实验步骤

（1）从实验场地的某一水准点出发，创建一条闭合水准路线，如图 2-1 所示，其长度以安置 4~5 个测站为宜，视线长度以 50 m 左右为宜。立尺点可以选择有凸出点的固定地物点或其他标记点，松软场地需安放尺垫，防止水准尺下沉。

（2）如图 2-1 所示，在起点 A（某一水准点）与第一个转点 B 的中间（前、后视的距离大致相等，采用目估或步测确定）安置并粗平水准仪。观测者按下列顺序观测：

① 后视立于水准点 A 的水准尺，精平，读数；
② 前视立于第一个转点 B 的水准尺，读数；
③ 改变水准仪的高度 10 cm 以上，重新安置水准仪；
④ 前视立于第一个转点 B 的水准尺，精平，读数；
⑤ 后视立于水准点 A 的水准尺，读数。

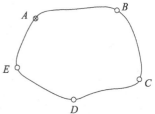

图 2-1　闭合水准路线

（3）测量记录时，观测者每次读数后，记录者应重复一遍读数，以便给观测者校对，确认无误后，当场记下。后视、前视观测完毕，应当场计算高差，并进行测站检核。

（4）依次设站，用相同的方法进行观测，直至回到出发点。

（5）全路线施测完毕，应进行线路检核，计算前视读数之和、后视读数之和、高差之和。若后视读数之和减前视读数之和等于高差之和，则说明计算过程无误。

2.4 注意事项

（1）在水准仪瞄准、读数时，水准尺必须立直。观测者从望远镜中根据纵丝与水准尺的相对位置可以判断尺子是否左右倾斜，立尺手可通过观测方向与伸展出去的立尺手臂呈"L"形来辅助判断尺子是否前后倾斜。

（2）在每个测站，两次仪高法测得的两个高差值之差不应大于 5 mm，否则该测站应重测。

（3）在每个测站，通过上述测站检核后才能搬站；仪器未搬迁时，前、后视水准尺的立尺点若放有尺垫，则均不得移动。若仪器搬迁了，则说明已通过测站检核，后视的立尺手才能携水准尺和尺垫前进至另一测站；前视的立尺手仍不得移动尺垫，只是将尺面转向，由前视转变为后视。

（4）闭合线路的高差闭合差容许值应满足以下规定：山地测量不应大于 $\pm 6\sqrt{n}$ mm，其中 n 为测站数；平地测量不应大于 $\pm 20\sqrt{L}$ mm，其中 L 为水准路线长度，单位为 km。

（5）将本次实验数据填写到表 2-1 中，完成相应的计算后，将其作为实验成果交给指导老师。

表 2-1　工程水准测量记录

班级：_____　组号：_____　日期：_____　观测者：_____

仪器编号：_____　记录者：_____

测站	点号	水准尺读数/mm		高差 h/m	平均高差/m	平差后的高差/m	高程 H/m	备注
		后视	前视					
检核计算 $\sum$								

实验 3　四等水准测量

3.1　实验目的

（1）掌握四等水准测量的观测和记录方法。
（2）熟悉四等水准测量的工作组织和一般规定，熟悉四等水准测量的主要技术要求，掌握测站和线路的检核方法。

3.2　实验安排和设备

（1）实验安排：实验课时为 2 学时；实验小组由 2~3 人组成。
（2）实验设备（以小组为单位）：水准仪 1 台，三脚架 1 副，双面水准尺 2 支，尺垫 2 个，记录板 1 块。学生自备计算器、铅笔、计算用纸。

3.3　实验方法和步骤

（1）创建一条闭合或附合水准路线，其长度以安置 4~6 个测站为宜。
（2）四等水准测量的观测步骤如下（以 DSZ3 型自动安平水准仪为例）：
① 使用圆水准器粗略整平仪器；
② 照准后视尺黑面，整平后读取上丝、下丝、中丝读数；
③ 照准后视尺红面，读取中丝读数；
④ 照准前视尺黑面，读取上丝、下丝读数，整平后读取中丝读数；
⑤ 照准前视尺红面，整平后读取中丝读数。
（3）按四等水准测量观测及记录的要求记录相关数据，记录完毕随即计算，计算结果符合各项限差后方可迁站。其中，后视距离 = 100×（后视上丝读数-后视下丝读数），前视距离 = 100×（前视上丝读数-前视下丝读数）。
（4）依次设站，并采用相同的方法施测其他各站。
（5）施测完毕后，计算闭合差、各站高差改正数及各待定点的高程。

3.4　四等水准测量的技术要求

四等水准测量的技术要求见表 3-1。

表 3-1　四等水准测量的技术要求

等级	标准视线长度/m	前、后视距差/m	前、后视距累计差/m	红、黑面读数差/mm	红、黑面高差之差/mm
四	100	5.0	10.0	3.0	5.0

3.5 检核

高差检核公式如下：

黑面：$\sum 后视 - \sum 前视 = \sum (后视 - 前视)$

红面：$\sum 后视 - \sum 前视 = \sum (后视 - 前视)$

视距差检核公式如下：

$\sum 本页后视距 - \sum 本页前视距 = 本页末站视距累计差 - 前页末站视距累计差$

3.6 注意事项

（1）四等水准测量有严格的技术规定，要求达到较高的精度，其关键在于：前、后视距要大致相等（在限差以内）；从后视转为前视（或相反），不能再整平；水准尺应竖直，最好使用附有圆水准器的水准尺。

（2）各项记录应整洁、清晰，如有读错、记错，必须重测，严禁涂改。

（3）每个测站上的记录、计算结果经检查全部合格后，才可迁站。

（4）将本次实验数据填写到表3-2中，完成相应计算后，作为实验的成果交给指导老师。

班级：_____ 组号：_____ 日期：_____ 观测者：_____ 仪器编号：_____ 记录者：_____

表 3-2 四等水准测量记录

序号	点名	读数				后视中丝/m	前视中丝/m	高差/m	高差改正值/m	改正后的高差/m	高程/m
		上丝(UL)/m 下丝(LL)/m	视距($UL-LL$)×100/m	视距差/m	视距累计差/m						
检核计算 Σ		—		—							

注：表中"—"表示无须填写任何数据。

实验 4　水准仪的检验与校正

4.1　实验目的

（1）了解水准仪各轴线之间应满足的条件。
（2）掌握水准仪的检验与校正方法。

4.2　实验安排和设备

（1）实验安排：实验课时为 2 学时；实验小组由 4 人组成，其中，1 人观测、检校，1 人记录，2 人立尺。

（2）实验设备（以小组为单位）：DSZ3 型自动安平水准仪 1 台，水准尺 2 支，尺垫 2 个，小螺丝刀 1 把，校正针 1 根，记录板 1 块。

4.3　实验方法和步骤

4.3.1　水准仪的检验、校正原理

DSZ3 型自动安平水准仪的轴线及其位置如图 4-1 所示。

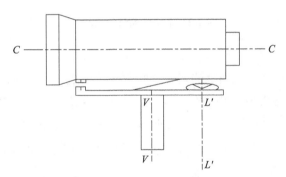

CC—水准仪视准轴；VV—仪器竖轴；$L'L'$—圆水准器轴。
图 4-1　DSZ3 型自动安平水准仪的轴线及其位置示意

水准测量时，水准仪必须提供水平视线。视线是否水平是根据圆水准器气泡是否居中来判断的。因此，水准仪的视准轴 CC 必须垂直于圆水准器轴 $L'L'$，这是水准仪应满足的主要条件。此外，水准仪还应满足以下两个条件：① 圆水准器轴平行于仪器竖轴（即 $L'L'//VV$）；② 十字丝横丝垂直于仪器竖轴。

4.3.2　水准仪的检验、校正

1. 圆水准器轴 $L'L'$ 的检验及校正

转动脚螺旋使圆水准器气泡居中，将照准部绕仪器竖轴 VV 旋转 180°后，若气泡仍

居中，则说明满足 $L'L'//VV$ 条件，否则需要校正。校正的方法：先稍松圆水准器底部中央的固定螺丝，再拨动圆水准器的校正螺丝，使气泡返回偏移量的一半，然后转动脚螺旋使气泡居中。如此反复检校几次，直至水准仪转至任何方向圆水准器气泡都不偏离中央，最后旋紧固定螺丝。

2. 十字丝的检验及校正（见图 4-2）

（1）检验方法：置平水准仪，以十字丝横丝的一端瞄准某一标志点 P，旋转水平微动螺旋，若点 P 始终在十字丝横丝上移动，如图 4-2a，b 所示，则说明十字丝横丝垂直于仪器竖轴；若标志点 P 移动的轨迹偏离了十字丝横丝，如图 4-2c，d 所示，则表明十字丝横丝不垂直于仪器竖轴，需要校正。

（2）校正方法：旋下十字丝分划板护罩，用小螺丝刀松开十字丝外环固定螺丝，微微转动外环，在水平方向微动时，使标志点 P 始终在横丝上移动，最后旋紧各种螺丝。

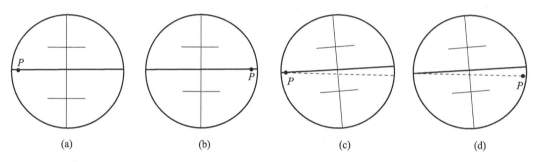

图 4-2　十字丝的检验及校正

3. 视准轴的检验及校正

如图 4-3 所示，在平坦地面上选定 4 个间距相等的点，相邻两点的间距为 40 m，在点 1 和点 2 处打木桩或安放尺垫，在点 A、点 B 两处竖立水准尺。先将水准仪安置于点 1，如图 4-3a 所示，整平仪器后，分别读取点 A、点 B 处水准尺的读数 r_A 和 r_B。然后将水准仪搬到点 2，如图 4-3b 所示，整平后分别读取点 A 和点 B 的水准尺读数 r'_A 和 r'_B。

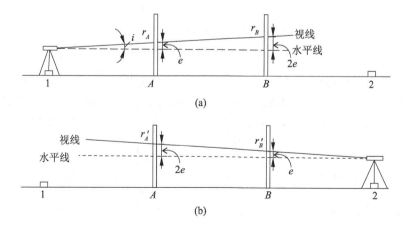

图 4-3　视准轴的检验及校正

如图 4-3a 所示，较长视距是较短视距的 2 倍，假设由视线不水平引起的点 A 处水准尺的读数误差为 e，那么点 B 处水准尺的读数误差为 2e。理论上，两次测得的点 A 和点 B 之间的高差应该相等，所以有如下等式：

$$(r_B-2e)-(r_A-e)=(r'_B-e)-(r'_A-2e) \tag{4-1}$$

从而得到

$$e=\frac{r_B-r_A-r'_B+r'_A}{2} \tag{4-2}$$

视准轴与水平线的夹角 i 的计算公式如下（注：1 rad = 206 265″）：

$$i=\frac{e}{D}\times 206\ 265'' \tag{4-3}$$

式中：D——点 A 和点 B 之间的水平距离。

若 $i<20''$，则水准仪不必校正，否则应校正。

水准管轴平行于视准轴的校正方法：卸下十字丝分划板外罩，用校正针拨动十字丝环的上、下两个校正螺丝，移动横丝，使对准点 A 的水准尺的正确读数为 r_A-e。该操作应反复进行，以确保校正成功。

4．补偿棱镜功能的检验

瞄准水准尺并读数，用手轻击三脚架的架脚，可看到十字丝产生震动。如果十字丝很快能稳定下来，并且横丝仍瞄准原来的读数，那么说明水准仪补偿棱镜功能正常。

4.4 注意事项

（1）必须按照实验步骤的顺序进行检验、校正，不能随意颠倒顺序。

（2）转动校正螺丝时，一次松紧的范围要小；校正完毕，校正螺丝应处于稍紧状态。

（3）实验结束后，完成表 4-1，并将其作为实验成果上交。

表 4-1 水准仪的检验与校正记录

班级：_____ 组号：_____ 日期：_____ 观测者：_____
仪器编号：_____ 记录者：_____

检验项目	检验过程	
	略图	观测数据及说明
圆水准器轴平行于竖轴		
十字丝横丝垂直于竖轴		
视准轴水平		$r_A =$ $r_A' =$ $r_B =$ $r_B' =$
		$e =$
		$i =$

实验 5　光学经纬仪的使用

经纬仪的操作

5.1　实验目的

（1）了解光学经纬仪的基本构造及主要部件的名称和作用。

（2）掌握光学经纬仪的基本操作方法——对中、整平、瞄准、读数。

5.2　实验安排和设备

（1）实验安排：实验课时为2学时；实验小组由4人组成，小组成员轮流操作仪器和记录读数。

（2）实验设备（以小组为单位）：光学经纬仪1台，记录板1块，测钎2支。

（3）指导老师安置觇牌若干块，将其作为各实验小组练习瞄准的目标。

5.3　实验方法和步骤

5.3.1　光学经纬仪的构造

光学经纬仪的构造如图5-1所示。

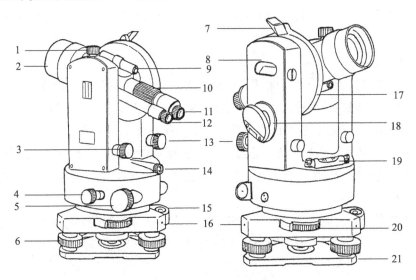

1—望远镜制动螺旋；2—望远镜物镜；3—望远镜微动螺旋；4—水平制动螺旋；5—水平微动螺旋；
6—脚螺旋；7—竖盘水准管观察镜；8—竖盘水准管；9—瞄准器；10—物镜调焦环；11—望远镜目镜；
12—度盘读数镜；13—竖盘水准管微动螺旋；14—光学对中器；15—圆水准器；16—基座；17—竖盘；
18—度盘照明镜；19—照准部水准管；20—水平度盘位置变换轮；21—基座底板。

图 5-1　光学经纬仪的构造示意

5.3.2 光学经纬仪的架设

(1) 将三脚架安置于地面标志的上方，展开三脚架腿以适应观测者的高度。观测者从三脚架的位置后退几步，检查三脚架头的中心是否在地面标志的垂线上，三脚架头尽可能水平。

(2) 从仪器箱内取出光学经纬仪，将其牢固地连结到三脚架头上，调节 3 个脚螺旋，使其高度大致相同。

5.3.3 光学经纬仪的基本操作

1. 对中

(1) 使用垂球粗对中

首先在连接螺旋下方挂一个垂球，调整垂球线的长度，使垂球略高于地面标志，然后进行对中。如果垂球偏离地面标志较大，可移动三脚架腿使垂球大致对准地面标志，然后将三脚架腿稳固地插入地面。最后，稍松开中心连结螺旋，并稍微移动三脚架头上的仪器使垂球对准地面标志的正上方。垂球粗对中的误差范围为 0~5 mm。

(2) 使用光学对中器精确对中

观测者略松开中心连结螺旋，稍微移动三脚架头上的仪器，观察光学对中器，直至地面标志影像与参考标志（十字丝或一个圆圈）重合。由于三脚架的顶面不完全水平，因此仪器必须平移。如果平移的同时旋转，那么光学对中器的视线不再是垂直的。平移操作后，必须检查仪器是否整平与对中。

2. 整平

经纬仪整平的目的是使水平度盘处于水平面。整平包括粗平和精平两个步骤。

粗平：调整架腿的长度使圆水准器气泡居中，其操作步骤详见实验 1。

精平：如图 5-2 所示，旋转脚螺旋使水准管气泡居中，其步骤如下。

① 旋转经纬仪使水准管平行于任意一对脚螺旋的连线，旋转这对脚螺旋使水准管气泡居中。脚螺旋要以相反的方向同时旋转，注意气泡的移动方向应与左手大拇指的移动方向一致（见图 5-2a）。

② 仪器旋转 90°，然后转动第 3 个脚螺旋，使水准管气泡居中（见图 5-2b）。

重复上述操作，直到仪器旋转至任意位置时，整平、对中都满足要求。

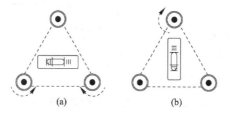

图 5-2 光学经纬仪的精平

3. 瞄准

松开照准部的水平制动螺旋，用望远镜上的瞄准器对准目标（测钎或觇牌），旋转望远镜制动螺旋和水平制动螺旋。调节目镜调焦螺旋，使十字丝清晰可见；调节物镜调焦螺旋，使目标像清晰；消除视差（与水准仪的视差消除操作相同）。旋转望远镜微动螺旋，使目标像的高度适中；转动水平微动螺旋，使目标像被十字丝的单根纵丝平分，或被两根纵丝夹在中央，完成瞄准。

4. 读数

首先调整度盘照明镜的位置，使读数视窗的亮度适中；然后调焦度盘读数镜的目镜，使度盘的分划清晰。光学经纬仪的读数视窗如图 5-3 所示，标明"水平"或"H"的为水平度盘读数；标明"竖直"或"V"的为垂直度盘读数。度盘的分微尺读数估读至 0.1′，并需将数值化为秒数。图 5-3 所示水平度盘的读数为 115°58′00″，垂直度盘的读数为 79°04′30″。

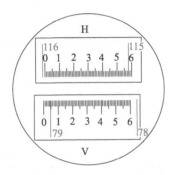

图 5-3 光学经纬仪的读数视窗

5. 其他练习

（1）盘左、盘右进行观测：松开望远镜制动螺旋，纵转望远镜，使其从盘左转为盘右（或相反），进行瞄准目标和读数练习。

（2）改变水平度盘的位置：旋紧水平制动螺旋，打开保护盖，转动水平度盘位置变换轮，从度盘读数镜中观察水平度盘读数的变化情况，并对准某一整数读数，例如，0°00′00″、90°00′00″等，之后盖好保护盖。

5.4 注意事项

（1）光学经纬仪对中时，应使三脚架的架头大致水平，否则会导致仪器整平困难。

（2）光学经纬仪整平时，应检查在各个方向平盘水准管气泡是否居中，其偏差应在规定的范围以内。

（3）瞄准目标时，必须先消除视差。

（4）完成表 5-1，并将其作为实验成果上交。

表 5-1 水平度盘读数练习记录

班级：_____　　组号：_____　　日期：_____　　观测者：_____

仪器编号：_____　　记录者：_____

测站点号	目标点号	竖盘位置	水平度盘读数			备注
			°	′	″	
		L				
		R				
		L				
		R				
		L				
		R				
		L				
		R				
		L				
		R				
		L				
		R				

实验 6　水平角观测

6.1　实验目的

掌握用经纬仪测回法观测水平角的操作、记录和计算方法。

6.2　实验安排和设备

（1）实验安排：实验课数为 2 学时；实验小组由 3 人组成，小组成员轮流观测和记录。

（2）实验设备（以小组为单位）：经纬仪 1 台，记录板 1 块，测钎 1~2 支。

6.3　实验方法和步骤

如图 6-1 所示，在测站点 B 处安置光学经纬仪，选择点 A 和点 C 作为目标，用测回法观测水平角 β 的方法与步骤如下：

图 6-1　测回法观测水平角

1. **盘左进行上半测回观测**

（1）光学经纬仪置于盘左（face left，FL）位置，转动设置螺旋，设置水平度盘的读数为零或在零附近。然后按一下水平度盘设置螺旋，使照准部与水平度盘分离（具体操作取决于所用的仪器）。

（2）松开望远镜制动螺旋，用瞄准器瞄准目标点 A，然后固定制动螺旋。转动望远镜微动螺旋，使望远镜的纵丝精确对准目标点 A（通常瞄准目标的底部）。必须注意消除视差，最后记下水平度盘读数 A_1。

（3）顺时针转动照准部，瞄准目标点 C，记下水平度盘读数 C_1，则点 A 与点 C 之间的水平角为 $\beta_1=C_1-A_1$。

2. **盘右进行下半测回观测**

倒转望远镜（望远镜绕横轴旋转 180°，照准部绕竖轴旋转 180°）使经纬仪处于盘右（face right，FR）位置，瞄准右方目标点 C，记录水平度盘读数 C_2。然后逆时针转动照准部，瞄准目标点 A，记录水平度盘读数 A_2。两读数之差即为下半测回角 β_2（即 $\beta_2=C_2-A_2$）。上、下两个半测回构成一测回。如果 β_1 与 β_2 的差值不大于 42″，那么取 β_1 与 β_2 的平均值作为一测回的最后角值，即 $\beta=(\beta_1+\beta_2)/2$。

当测角精度要求较高时，需观测数个测回。为了减少度盘分划不均的影响，在每一测回观测之后，要根据测回数 n，先将度盘读数设为 $180°/n$，再开始下一测回的观测。因此，如果需观测 2 个测回，那么在开始第二个测回时，应瞄准目标点 A，设置水平度盘的读数为 90°。

每个测站至少观测 2 个测回，以便于角度计算时检测误差。由于每个测回都是独立观测的,因此在计算与比较之后,只有两个测回的误差在规定范围时,才能搬迁仪器与三脚架。应该注意的是,每次瞄准目标时,都要使用纵丝中央的同一位置,以便减少瞄准的误差。

6.4　注意事项

(1) 安置经纬仪时,与地面点的对中误差应小于 2 mm。

(2) 瞄准目标时,应尽量瞄准目标底部,以减小由目标倾斜引起的水平角观测误差。

(3) 观测过程中,若发现水准管气泡偏移超过 2 格,应重新整平仪器,并重新观测该测回。

(4) 每人至少应独立进行一测回的水平角观测,将观测结果填入表 6-1 中,并将其作为实验成果上交。

表 6-1 水平角观测（测回法）记录

班级：_____ 组号：_____ 日期：_____ 观测者：_____

仪器编号：_____ 记录者：_____

测站点号	目标点号	竖盘位置	水平盘读数/ (° ′ ″)	水平角值/ (° ′ ″)		备注
				半测回	一测回	
		L				
		R				
		L				
		R				
		L				
		R				
		L				
		R				
		L				
		R				
		L				
		R				
画草图						

实验 7　竖直角测量及竖盘指标差检校

7.1　实验目的

（1）认识经纬仪竖盘的构造，了解其注记形式，了解竖盘指标差与竖盘水准管之间的关系。

（2）掌握竖直角的测量方法。

（3）掌握竖盘指标差的检验和校正方法。

7.2　实验安排和设备

（1）实验安排：实验课时为 2 学时；实验小组由 2 人组成。

（2）实验设备（以小组为单位）：经纬仪 1 台，记录板 1 块。

（3）指导老师安置觇牌若干块，作为各实验小组练习瞄准的目标。

7.3　实验方法和步骤

7.3.1　竖直角测量

（1）在指定的测站上安置好仪器，进行对中、整平，转动望远镜，从读数镜中观察竖盘读数的变化，确定竖盘的注记形式，并在记录表中写出竖直角及竖盘指标差的计算公式。

（2）选定某一觇牌或其他明显标志作为目标。盘左，瞄准目标（用十字丝的中横丝切于目标顶部），转动竖盘水准管微动螺旋，使竖盘水准管气泡居中后，读取竖盘读数 L，并计算盘左半测回竖直角值 $\alpha_{左}(\alpha_{左}=90°-L)$。

（3）盘右，做同样的观测，记录竖盘读数 R，并计算盘右半测回竖直角值 $\alpha_{右}(\alpha_{右}=R-270°)$。

（4）计算指标差 $\bar{\chi}$ 及一测回竖直角 α。

$$\begin{cases} \bar{\chi} = \dfrac{1}{2}(\alpha_{右}-\alpha_{左}) \\ \alpha = \dfrac{1}{2}(\alpha_{左}+\alpha_{右}) \end{cases} \quad (7\text{-}1)$$

（5）每人应至少对同一目标观测 2 个测回，或对两个不同的目标各观测一测回。指标差对于某一仪器应为一个常数，因此，各测回测得的指标差之差不应大于 20″。

7.3.2　竖盘指标差的检验和校正

检验各测回观测计算所得的指标差之差是否超限，剔除离群值，取其平均数作为该仪器的竖盘指标差$\bar{x}$。如果$\bar{x}$的绝对值大于60″，则需要进行指标差的校正。

竖盘指标差的检验与校正的目的是当视线水平以及当指标水准管气泡居中时，确保竖盘的读数是90°或90°的倍数（依仪器的类型而定）。

1. 竖盘指标差的检验

假设仪器的竖盘分划是顺时针增加的，并且经纬仪置于盘左，视线水平时，竖盘读数为90°。竖盘指标差的检验步骤如下：

（1）使经纬仪置于盘左（FL），仔细整平仪器。

（2）用望远镜十字丝的横丝精确地瞄准一个精细的目标点。

（3）使指标水准管气泡居中，并读取竖盘读数L。

（4）纵转望远镜，使经纬仪处于盘右（FR），重复步骤（2），再调整指标水准管气泡使其居中，并读取竖盘读数R。

例如，检验结果中，盘左竖盘读数$L=78°18′18″$，盘右竖盘读数$R=281°42′00″$。按公式（7-1）求得

$$\bar{x}=[(281°42′00″-270°)-(90°-78°18′18″)]/2 = 9″$$

2. 竖盘指标差的校正

经纬仪位于盘右，望远镜仍瞄准目标点。

（1）当经纬仪位于盘右时，竖盘的正确读数应为$R-\bar{x}$，即$R-\bar{x}=281°42′00″-9″=281°41′51″$。旋转指标水准管的水平螺旋使竖盘指标对准这个读数，这会导致指标水准管气泡偏离中心。

（2）调节指标水准管的校正螺丝使指标水准管气泡居中。

注意：对于自动垂直指标的经纬仪，应查阅厂家仪器手册获得正确的校正方法。

7.4 注意事项

（1）观测目标时，先调清楚十字丝，然后消除视差，尽量用十字丝的交点来瞄准目标，每次读数时都要使指标水准管气泡居中。

（2）计算竖直角时要注意其正、负号。

（3）实验结束后填写表7-1，并将其作为本次的实验成果上交。

表 7-1 竖直角观测记录

班级：_____ 组号：_____ 日期：_____ 观测者：_____

仪器编号：_____ 记录者：_____

测站	目标点号	竖盘位置	竖盘读数			竖直角						指标差
						半测回			一测回			
			°	′	″	°	′	″	°	′	″	″
		FL										
		FR										
		FL										
		FR										
		FL										
		FR										
		FL										
		FR										
		FL										
		FR										
		FL										
		FR										
		FL										
		FR										
		FL										
		FR										
		FL										
		FR										
		FL										
		FR										
		FL										
		FR										
		FL										
		FR										
竖直角及指标差计算公式												

实验 8　经纬仪的检验和校正

8.1　实验目的

（1）了解经纬仪的主要轴线应满足的几何条件。
（2）掌握经纬仪检验和校正的基本方法。

8.2　实验安排和设备

（1）实验安排：实验课时为 2 学时；实验小组由 3 人组成。
（2）实验设备（以小组为单位）：经纬仪 1 台，测钎 3 支，水准尺 1 支，记录板 1 块，校正针 1 根。

8.3　实验原理

为了能正确地测出水平角和竖直角，经纬仪要能够精确地安置在测站点上，且能使竖轴与地面控制点位于同一铅垂线上；当望远镜绕横轴旋转时，能够形成一个铅垂面；当望远镜水平时，竖盘读数应为 90°或 270°。

为满足上述要求，经纬仪应满足以下几何条件：
（1）照准部的水准管轴垂直于竖轴；
（2）十字丝纵丝垂直于横轴；
（3）望远镜视准轴垂直于横轴；
（4）横轴垂直于竖轴；
（5）光学对中器的视准轴与竖轴的旋转中心线重合；
（6）竖盘读数指标处于正确的位置。
本实验的内容即检验经纬仪是否满足以上几何条件。

8.4　实验方法和步骤

8.4.1　照准部水准管轴垂直于竖轴的检验和校正

1. 检验要求
（1）每人独立检验一次并做好记录。
（2）根据两次的检验结果确定仪器是否满足几何条件。若两次检验中水准管气泡的偏移量均小于半格，即基本满足条件。
2. 检验步骤
初步整平仪器，转动照准部使平盘水准管平行于一对脚螺旋的连线，转动这对脚螺旋使水准管气泡严格居中。再将照准部旋转 180°，如果气泡仍然居中，则说明平盘水准管轴垂直于竖轴，否则需要校正。

3. 校正步骤

校正时，拨动水准管校正螺丝，使气泡返回偏移量的一半，旋转脚螺旋使气泡返回偏移量的另一半。反复检校，直至水准管轴旋转至任何位置时水准管气泡偏移量都在一格以内。

8.4.2 视准轴垂直于横轴的检验和校正

1. 检验要求

（1）检验应每人分别独立进行，既可使用横尺读数法，也可使用 1/4 法。

（2）分别记录两次倒镜后在横尺上的读数，计算读数差。

（3）若两次读数差的差值小于 3 mm，则可取其平均值作为该仪器的视准轴与横轴的关系资料，否则应重新检验一次。

2. 检校步骤

（1）方法一

如图 8-1 所示，盘左瞄准远处大致与仪器同高度的目标点 A，读取水平度盘读数 $\alpha_{左}$；再将盘右瞄准目标点 A，读取水平度盘读数 $\alpha_{右}$。若 $\alpha_{右}=\alpha_{左}\pm 180°$，则说明视准轴垂直于横轴，否则需要校正。

校正时，先计算盘右瞄准目标点 A 时的应有读数，即

$$\alpha'_{右}=\frac{1}{2}[\alpha_{右}+(\alpha_{左}\pm 180°)]$$

转动水平微动螺旋，使水平度盘读数为 $\alpha'_{右}$，再拨动十字丝左、右一对校正螺丝，使十字丝纵丝瞄准目标点 A。如此反复检校，直至盘左、盘右读数加减 180° 后的差值（$2c$ 值）小于 60″。最后旋上十字丝环护罩。

（2）方法二

如图 8-1 所示，在平坦地面上选择相距约 100 m 的 A，B 两点，将经纬仪置于点 A 和点 B 连线的中点 O 处，在点 A 处放一块大致与经纬仪同高的觇牌，在点 B 大致与经纬仪同高处水平地放一把分划尺，方向与 OB 垂直。盘左瞄准点 A 处的觇牌（水平制动螺旋

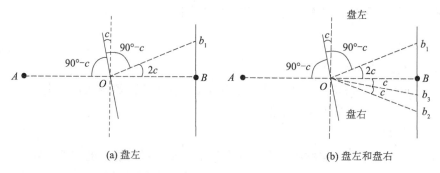

图 8-1 视准轴垂直于横轴的检验与校正原理

应制紧），倒转望远镜，在点 B 处的尺上读数，记为 b_1；盘右用同样的方法，在尺上读数，记为 b_2。若 $b_2=b_1$，则说明视准轴垂直于横轴，否则需要校正。

拨动十字丝左、右一对校正螺丝，使纵丝瞄准尺上读数为 b_3。反复检校，直至满足要求。

$$b_3 = b_1 + \frac{3}{4}(b_2 - b_1)$$

校正要求：当两读数差的平均值 ≥ 8 mm 时应校正。

8.4.3 十字丝的检验和校正

1. 十字丝的检验

（1）检验要求

① 每人分别独立检验一次，记录纵丝对一个固定点由一端到另一端的偏离长度（估计读数，以 mm 计）。

② 以两次检验的近似结果描述仪器的状态。

（2）检验步骤

用望远镜中十字丝交点瞄准一个目标点，旋转望远镜微动螺旋，使纵丝上下移动，如果目标点始终不离开纵丝，则说明纵丝垂直于横轴，否则需要校正。

2. 十字丝的校正

（1）校正要求

① 若两次的检验结果中纵丝一端均偏离固定点 2 mm 以上，则应校正（见图 8-2a）。

② 校正后再检验，纵丝不应偏离固定点。

（2）校正步骤

校正时，旋下十字丝环护罩，用小螺丝刀松开十字丝外环的 4 个固定螺丝（见图 8-2b），转动十字丝环，使望远镜上下微动时点 P 始终在纵丝上移动（见图 8-2c），最后旋紧十字丝外环的固定螺丝。

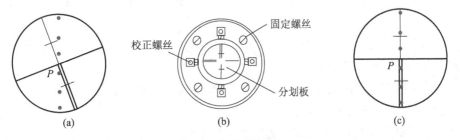

图 8-2 十字丝的校正

十字丝本身纵丝与横丝是相互垂直的，设置纵丝垂直也就设置了横丝水平。

视准轴和十字丝的检验应分别进行，两者的校正可以同时进行，直到每一项都获得令人满意的结果。

8.4.4 光学对中器的检验和校正

1. 检验要求

光学对中器的视准轴经棱镜折射后应与仪器的竖轴相重合。若不重合，则照准部旋转时将产生对中误差。光学对中器的结构示意如图 8-3 所示。

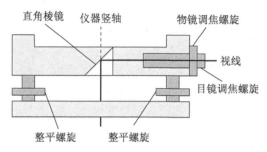

图 8-3 光学对中器的结构示意

光学对中器的检验要求如下：
（1）每人独立检验一次，并做好记录。
（2）根据两个人的检验结果确定仪器是否满足条件。

2. 检校步骤

首先把仪器架好，并整平。然后在仪器正下方的地面固定一张纸，在光学对中器轴线与纸的相交处做第一个标志。水平旋转照准部 180°，在光学对中器轴线与纸的相交处做第二个标志。若两个标志重合，则光学对中器不必调整；若不重合，则两个标志的中点就是光学对中器轴的正确位置。查阅仪器使用手册，调节光学对中器的十字环或物镜，使光学对中器轴与仪器竖轴重合。

8.5 注意事项

（1）检验和校正过程中应爱护仪器，不得随意拨动仪器的各个螺丝。
（2）按实验步骤进行各项检验和校正操作，顺序不能颠倒，检验数据正确无误后才能进行校正。
（3）需要校正时，应向指导老师说明仪器的关系资料和应当采用的校正方法，征得指导老师的同意后方可进行校正。
（4）校正应在老师的指导下进行，检验和校正应反复进行，直至满足要求。校正结束时，各校正螺丝应处于稍紧状态。
（5）认真填写表 8-1，并将其作为实验成果上交。

表 8-1 经纬仪的检验和校正记录

班级：_____ 组号：_____ 日期：_____ 观测者：_____

仪器编号：_____ 记录者：_____

检验项目	检验过程	
	略图	观测数据及说明
水准管轴⊥竖轴		
圆水准轴∥竖轴		
横丝⊥竖轴		
视准轴⊥横轴		
横轴⊥竖轴		

实验 9　钢尺量距

9.1　实验目的与要求

9.1.1　实验目的

（1）掌握钢尺量距的方法，以及量距的计算方法。
（2）掌握用标杆定直线的方法。

9.1.2　实验要求

钢尺量距的相对误差小于 1/2 000。

9.2　实验安排和设备

（1）实验安排：实验课时为 2 学时；实验小组由 3 人组成。
（2）实验设备（以小组为单位）：钢卷尺（30 m）1 个，花杆 3 根，测钎 4 支，记录板 1 块。

9.3　实验方法和步骤

钢尺量距是距离测量的基本方法。丈量前，先将待测距离的两个端点用木桩（桩顶钉一个小钉）标识出来，清除直线上的障碍物后，一般由两人分别在待测距离的两个端点边定线边丈量，具体的操作方法下。

（1）如图 9-1 所示，量距时，先在 A，B 两点上竖立测杆（或测钎），标定直线方向，后尺手持钢尺的零端位于点 A，前尺手持钢尺的末端并携带一束测钎，沿 AB 方向前进，至一尺段长处停下，两人都蹲下。

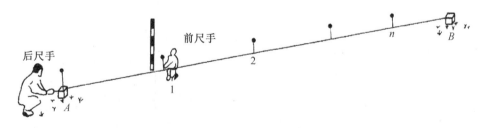

图 9-1　平坦地面上的量距方法

（2）后尺手以手势指挥前尺手将钢尺拉在 AB 直线方向上；两人同时将钢尺拉紧、拉平、拉稳后，前尺手喊"预备"，后尺手将钢尺零点准确对准点 A，并喊"好"，前尺手随即将测钎对准钢尺末端刻划并竖直地插入地面（在坚硬地面处，可用铅笔在地面划线作为标记），得到点 1。这样便完成了第一尺段的丈量工作。

(3) 后尺手与前尺手共同举尺前进，后尺手走到点 1 时，即喊"停"，同上述方法丈量第二尺段，然后后尺手拔起点 1 上的测钎带在身上。如此继续丈量下去，直至最后量出不足一整尺的余长 q，则点 A 和点 B 之间的水平距离为

$$D = nl + q \tag{9-1}$$

式中：n——整尺段数（即在点 A 和点 B 之间后尺手所拔测钎数）；

l——钢尺长度，m；

q——不足一整尺的余长，m。

为了防止丈量错误和提高精度，一般还应由点 B 至点 A 进行返测，返测时应重新进行定线。取往、返测距离的平均值作为 AB 最终的水平距离。

$$D_{av} = \frac{1}{2}(D_f + D_b) \tag{9-2}$$

式中：D_{av}——往、返测距离的平均值，m；

D_f——往测的距离，m；

D_b——返测的距离，m。

量距精度通常用相对误差 K 来衡量，相对误差 K 可化为分子为 1 的分数形式，即

$$K = \frac{|D_f - D_b|}{D_{av}} = \frac{1}{D_{av}/|D_f - D_b|} \tag{9-3}$$

9.4 注意事项

（1）返测须重新定线；测钎要插直，测钎数不要记错；每一段的定线要准确，使钢尺在直线内丈量。

（2）丈量距离时要注意钢尺零点和终点的位置，以及米、分米的注记，以防读错。

（3）应注意拉平钢尺，力求保持拉力一致。若遇斜坡或坑洼不平的地带，则用花杆或吊垂球直接丈量出水平距离。这时要特别注意将钢尺端点准确地投影到地面上。

（4）为维护钢尺，应做到"四不"：不扭、不折、不压、不拖。钢尺用完后要擦拭干净才可卷入尺壳内。

（5）完成表 9-1，并将其作为实验成果上交。

表 9-1　钢尺量距记录

班级：_____　　组号：_____　　日期：_____　　观测者：_____

仪器编号：_____　　记录者：_____

线段名称	观测回数	整尺段数 n	余长 q/m	距离 D/m	平均距离 D_{av}/m	相对误差 K	温度/℃
	往测						
	返测						
	往测						
	返测						
	往测						
	返测						
	往测						
	返测						
	往测						
	返测						
	往测						
	返测						
	往测						
	返测						
	往测						
	返测						
	往测						
	返测						

实验10　全站仪的使用

全站仪的使用

10.1　实验目的

（1）了解 2″级电子全站仪的功能及其主要部件的名称和作用。

（2）掌握电子全站仪的基本操作方法，并练习水平角、竖直角和距离的测量等操作。

（3）熟悉数字化测图的坐标测定和施工放样的点位测设。

10.2　实验安排和设备

（1）实验安排：实验课时为4学时；实验小组由2~3人组成，小组成员轮流操作仪器，并做记录。

（2）实验设备（以小组为单位）：2″级电子全站仪1台，反光棱镜1个或者反射片5~6片，2 m钢卷尺1个。

10.3　实验方法和步骤

10.3.1　准备工作

（1）观测前先了解全站仪的结构。全站仪的结构示意如图10-1所示。

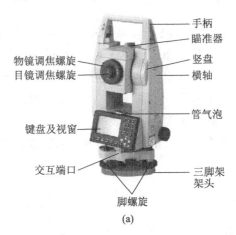

(a)

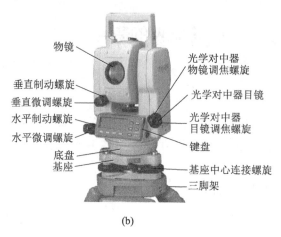

(b)

图 10-1　电子全站仪的结构示意

（2）安装电池，按【ON】键开机，仪器进行自检。旋松水平和垂直制动螺旋，使照准部在水平方向；将望远镜在垂直方向旋转一周，则水平度盘和垂直度盘指标设置完毕，视窗出现"测量模式"屏幕。

视窗下的【F1】—【F4】为功能键（又称"软键"），可实现全站仪的各种功能。应用全站仪的功能时，需要按照视窗内光标的指示，输入相应的数字或字母。另外，【Backspace】键可删除已输入（光标左边）的字符；【Esc】键可取消输入的数据内容。

10.3.2　全站仪的安置

将三脚架放在测站点的上方，使三脚架架头大致水平，将仪器放到三脚架架头上，一只手提着仪器的手柄，另一只手旋紧连接螺旋。移动三脚架，将光学对中器对准测站点，伸缩三脚架腿，使圆水准器气泡居中。转动照准部，使水准管平行于一对脚螺旋的连线，相对旋转脚螺旋，使水准管气泡居中；将照准部旋转90°，调节第三个脚螺旋使气泡居中。检查光学对中器的对中情况，必要时，略松连接螺旋，平移仪器使其精确对中，再检查水准管气泡的居中情况。重复以上步骤，直到仪器既对中又整平。

10.3.3　角度测量

1. 水平角测量

如图10-2所示，通常先在正镜（盘左）位置，从测站点 S 瞄准左目标 L 的觇牌，在测量模式菜单中按下置零按钮，使瞄准该方向的水平度盘的读数为 $0°00'00''$。转动照准部，瞄准右目标 R，假如水平度盘的读数为 $106°16'20''$，该值即为盘左位置测量的水平角 α。在倒镜（盘右）位置（先将照准部转动180°，再将望远镜绕横轴旋转180°，即可从

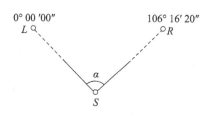

图 10-2　水平角测量

盘左位置变为盘右位置）采用同样的方法进行盘右位置的水平角测量。若两次测得的水平角的差值不大于42″，则取其平均值作为一测回该水平角α的值；若差值大于42″，则需重测该测回。

2. 竖直角测量

如果需要同时测定目标的竖直角，在角度测量模式下，可以在瞄准目标时的屏幕上读取天顶距。如图10-3所示，假设瞄准目标L时的读数（此读数为天顶距的值）为49°59′50″，此时的竖直角β为90°−49°59′50″=40°00′10″。

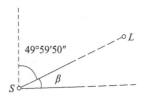

图 10-3　竖直角测量

在倒镜（盘右）位置采用同样的方法进行竖直角测量。此时，竖直角β=R−270°，其中R为盘右位置时的竖盘读数。若两次测得的竖直角的差值不大于42″，则取其平均值作为一测回竖直角β的值；若其差值大于42″，则需要测该测回。

10.3.4　距离测量

1. 参数设置

在测量前，先对测量模式、目标类型、棱镜常数、气温、气压等进行设置。测距参数的名称及含义如下：

（1）Mode（测量模式），其选项包括Fine "r"（重复精测）、Fine AVG（平均精测）、Fine "S"（单次精测）、Rapid "r"（重复粗测）、Rapid "S"（单次粗测）、Tracking（跟踪测距）。

（2）Reflector（目标类型），其选项包括Prism（棱镜）、Sheet（反射片）。

（3）PC（prism constant，棱镜常数），其数值按所用棱镜输入（以mm为单位），见各仪器说明书。

（4）Temp（气温），按测距当时所测气温输入（单位为℃）。

（5）Press（气压），按测距当时所测气压输入（单位为hPa）。

（6）PPM（气象改正百分比），仪器自动计算。

2. 测距步骤

瞄准具有反射棱镜或者贴有反射片的目标，在测量模式下，按下"距离测量"按钮，瞄准目标后，按下软键盘上的"测量"按钮，仪器就开始按设置好的测距参数进行距离测量。"滴"一声响后，屏幕显示斜距SD、垂距VD、平距HD的值。

若将测距模式设置为"单次精测"，则一次测量完成后，测量自动停止。若设置为"平均精测"，则显示各次测距值为S-1，S-2，…，S-9，测量完成后，显示距离的平均值。距离和角度的最新测量值自动保存在内存中，可以随时调阅，保存的数据在关机后即被清除。

10.3.5　坐标测量

全站仪的三维坐标（X，Y，Z）用于地形测量的数据采集。根据测站及后视点（定向点）的已知坐标（或方位角），通过距离和角度观测，测定观测点的三维坐标，并将其存储于内存中。

坐标测量的步骤如下：

1. 选取工作文件

可以选取或新建任何一个文件作为"当前工作文件"，用以记录本次测量结果。将光标移至要选取的文件名上（或输入新建文件的文件名），按回车键，则该文件即为当前测量的数据记录文件。

2. 输入测站数据

如图 10-4 所示，O 为测站点，A 为后视点，P 为前视点，OPQ 为观测的前进方向。在坐标测量模式下，输入下列各项数据：测站点坐标（N_0，E_0，Z_0）、仪器高（Inst. h.）、目标高（Tgt. h.）、后视点坐标（X_A，Y_A）。每输入一项数据后按回车键确定，测站数据记录在所选取的文件中。

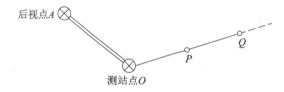

图 10-4　坐标测量示意

3. 测量三维坐标

完成测站相关数据的输入后，瞄准后视点 A 按回车键确认，其后显示的数值即为全站仪自动计算的测站点至后视点的方位角。回到"坐标测量界面"，瞄准目标点 P，通过斜距 S、天顶距 ZA 和目标方位角 HAR 的测量，可以确定目标点 P 的三维坐标（N_P，E_P，Z_P）。计算公式如下：

$$\begin{cases} N_P = N_0 + S\sin ZA\cos HAR \\ E_P = E_0 + S\sin ZA\sin HAR \\ Z_P = Z_0 + S\sin ZA + h_I - h_T \end{cases}$$

式中，h_I 为仪器高，h_T 为目标高。计算由仪器自动完成并显示于屏幕，存储于所选的工作文件中。

瞄准目标点 P 的棱镜后，在坐标测量模式下点击"测量"按钮，即显示目标点的三维坐标、距离和角度观测值。若继续测量，则把仪器搬至图 10-4 中的点 P，此时点 P 为测站点，点 O 为后视点，点 Q 为前视点，重复以上步骤，进行下一点的坐标测定。

10.3.6　放样

放样是在实地标记出已知设计坐标（或方向和距离）所指定的点。在放样测量中，根据棱镜所在点的水平角、距离或坐标，仪器能显示预先输入的待放样值与实测值之差，即

显示差值=实测值-放样值

根据差值，目标点的棱镜觇牌做相应的移动，使差值逐步小于容许值，最终精确找到设计坐标指定的点，达到放样的目的。放样测量通常在盘左位置进行。

1. 角度放样和距离放样

在测站上，根据某一基准方向转过的水平角和距离，即可测设待定点。

在测站上安置仪器后，照准后视点。按放样功能键，将基准方向设置为零。在角度距离放样时需输入下列两项数据：放样距离（S.O.dist）和放样角度（S.O.H.ang），每输完一项数据后按下回车键确认。移动棱镜或觇标，不断跟踪测量，其中，S-O-S在观测后会显示至观测点的距离，dHA（水平角偏移）显示棱镜所在位置至观测点的连线与初始方向线的水平角。在观测过程中，随着棱镜的移动，屏幕显示的放样数据也随之变化，逐渐接近直至精确找到待放样点。

2. 坐标放样

坐标放样是在实地标记出坐标值已知的点。假设测站点和后视点的坐标为已知，在输入待放样点的坐标后，仪器将自动计算出放样所需的方位角和平距并存储于内存中。利用角度放样和距离放样功能，可测设待放样点的位置。

在测量模式菜单中按下【S-O】软键，显示"放样测量菜单"屏幕。输入测站点的三维坐标（N_0，E_0，Z_0）、仪器高（Inst. h.）和目标高（Tgt. h.）。输完每项数据后按回车键，全部输完后按【OK】软键，回到"放样测量菜单"屏幕，此时建站完成。接着输入待放样点的三维坐标，此时，屏幕将显示目标点与当前位置的距离差和方位角差，如图10-5所示。

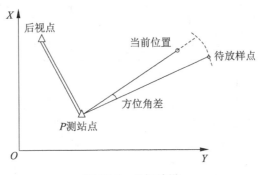

图 10-5 坐标放样

在观测过程中，随着棱镜的移动，屏幕显示的放样数据也随之变化，逐渐接近要找的待放样点。最后调节水平和竖直微调螺旋，直到屏幕显示的距离和角度都为0。此时，已精确找到待测点的位置，在对应的位置做好放样标记即可。

10.4 注意事项

（1）爱惜仪器，轻拿轻放，注意防潮。

（2）每人至少应完成一个测站的全部测量。

（3）将本次实验的观测数据记录于表10-1中，并作为实验成果上交。

表10-1 全站仪测量记录表

班级：_____ 组号：_____ 观测者：_____ 记录者：_____
镜高：_____ m 日期：_____ 仪器编号：_____

序号	点号	仪器高/m	转折角/(° ′ ″)	距离/m	坐标 x/m	坐标 y/m	坐标 z/m	坐标增量改正值 Δx/m	Δy/m	Δz/m	改正后的坐标 x/m	y/m	z/m
1	后视点			—				—	—	—	—	—	—
1	测站点												
1	前视点												
2	后视点			—				—	—	—	—	—	—
2	测站点												
2	前视点												
3	后视点			—				—	—	—	—	—	—
3	测站点												
3	前视点												
4	后视点			—				—	—	—	—	—	—
4	测站点												
4	前视点												
5	后视点			—				—	—	—	—	—	—
5	测站点												
5	前视点												
检核													

注：表中"—"表示无须填写任何数据。

实验 11　GPS 的认识与使用

GPS 的使用

11.1　实验目的

（1）掌握 GPS 仪器的基本操作以及控制点点位的选择方法。
（2）掌握 GPS 的施测、记录方法。

11.2　实验安排和设备

（1）实验安排：实验课时为 2 学时；实验小组由 2 人组成，其中，1 人操作仪器，1 人记录，轮流执行相应的实验操作。
（2）实验设备（以小组为单位）：GPS 1 台（本次实验以思拓力 S5Ⅱ为例），配套手簿 1 个，记录板 1 块。

11.3　实验方法和步骤

通常情况下，使用 GPS 测量的步骤如下：
（1）架设基准站。注意：连接 CORS 差分信号时不需要架设基准站。
（2）打开手簿中的 SurPad4.0 软件，手簿与基准站相连后，新建项目，设置坐标系统参数和基准站参数，使基准站发射差分信号。
（3）连接移动站，并设置移动站，使移动站接收到基准站的差分数据，并达到固定解。
（4）移动站到测区已知点上测量出固定解状态下已知点的原始 WGS-84 坐标，根据已知点的原始坐标和当地坐标求解出两个坐标系之间的转换参数，并应用转换参数。
（5）到另外一个已知点检查转换之后的当地坐标是否正确。
（6）开始作业（如果基准站重新开机或位置挪动，移动站需要进行基站平移、校准，利用标记点校正）。
（7）将数据文件导出为需要的格式。

11.3.1　准备工作

将手机 SIM 卡（中国移动、中国联通或者中国电信的 SIM 卡都可以，SIM 卡需联网）和存储卡安装到 GPS 仪器对应的卡槽里，安装电池，打开接收机开始接收卫星信号。采用该模式，实现基准站通过手机网络将数据发给移动站。

11.3.2　连接仪器及设置工作模式

开机后，执行"仪器"—"通讯设置"命令，仪器类型选择"RTK"，通信模式选择"蓝牙模式"，点击"搜索"按钮，在蓝牙设备列表中找到自己仪器的蓝牙名称，点击"连接"按钮,弹出连接进度框,进度完成后,表示测量手簿和仪器连接成功。

11.3.3 主机自检

自检功能的主要作用是判断接收机的各个模块是否工作正常。当S5Ⅱ的指示灯不亮或者有模块不能正常工作时,可使用自检功能来检测接收机。S5Ⅱ自检部分包括GNSS、电台、Wi-Fi、蓝牙和传感器5个部分。自检过程中会有语音播报检测结果。

执行"仪器"—"移动站模式"命令,数据链模式选择"手簿网络",设置好CORS服务器的IP地址和端口,获取并选择接入点,其他选项可以使用默认数值,点击"应用"按钮,工作模式即设置完毕。返回主界面,持稳仪器,可以从测量手簿查看是否得到固定解。

11.3.4 新建项目

运行SurPad4.0软件,执行"项目"—"项目管理"—"新建"命令,新建项目,在弹出的对话框中输入项目名称,选择坐标参数类型,其他为附加信息,可留空,点击"确定"按钮,跳转到坐标系统参数界面。在界面上点击"投影参数"按钮,在下拉列表里选择投影方式,一般选择"高斯投影"。设置中央子午线的参数,参数值可以直接输入,也可以通过手簿中软件自动计算并添加(这里是指手簿连上仪器,并且仪器已经锁定卫星的情况),如图11-1所示。

图 11-1 投影参数设置

11.3.5 校正

1. 基站校正

在固定解状态,执行"项目"—"测站校准"—"利用基站点校准"命令,输入已知坐标,设置当前基站坐标的天线参数,点击"计算"按钮得到校准参数。

2. 点校正(求转换参数)

第一次进入测量区域时,如果要将已知点的坐标与待测量的点相匹配,则需要进行点校正。

设置完仪器,待移动站达到固定解后就可以求参数或校准了,由于RTK测出来的原始数据是WGS-84坐标,所以必须通过参数求解,把仪器所测得的数据转成需要的坐标系下的坐标,比如国家80或北京54坐标。

具体做法:

将移动站架设到控制点上,立好对中杆并使气泡居中,使用"点测量"功能把控制点(4个左右)的原始WGS-84坐标采集下来。然后执行"工具"—"转换参数"命令,依次输入已知点的坐标(从坐标点库中选取或手动输入)和WGS-84椭球原始坐标

（获取当前 GPS 数据，从坐标点库中选取或手动输入），设置是否使用平面校正和高程校正，点击"确定"按钮，完成转换参数的输入。在转换参数界面点击"计算"按钮得到 GPS 参数报告。

11.3.6 测量

将移动站架设到控制点上，设置天线高度，对中整平，执行"测量"—"点测量/碎部测量"命令。以"地形点"为例，点击"点类型"按钮，选择地形点，再点击"设置"按钮记录地形点限制条件（例如，固定解，H：0.05；V：0.1；PDOP：3.0；延迟：5；平滑：1），点击右下角"采点"按钮或者使用手簿采点快捷键完成目标点的采集和保存。

11.3.7 数据导出

从手簿中导出数据，并拷贝至电脑中进行内业处理。

11.5 注意事项

（1）基准站附近应无大型遮挡物，无电磁波干扰（要求 200 m 内没有微波站、雷达站、手机信号站等，50 m 内无高压线）。

（2）至少有 2 个坐标已知点（已知点可以是任意坐标系下的坐标，已知点最好有 3 个或 3 个以上，以便检校已知点的正确性）。

（3）将本实验的观测数据记录在表 11-1 中，并作为实验成果上交。

表 11-1　GPS 测量记录

班级：_____　　组号：_____　　日期：_____　　观测者：_____

仪器编号：_____　　记录者：_____

经度：_____　　纬度：_____　　高度：_____　　温度：_____ ℃

压力：_____　　湿度：_____　　精度：_____　　初始倾斜高度：_____ m

最终倾斜高度：_____ m

历史古迹描述（含表面状况）：

潜在问题：

与古迹的联络情况：

意见：

点号	坐标		
	x	y	z

实验 12　建筑物轴线测设和高程测设

12.1　实验目的

（1）掌握建筑物轴线测设的基本方法。
（2）掌握建筑施工中高程测设的基本方法。

12.2　实验安排和设备

（1）实验安排：实验课时为 2 学时；实验小组由 3 人组成。
（2）实验设备（以小组为单位）：全站仪 1 台，水准仪 1 台，钢尺 1 把，标杆 1 支，水准尺 1 支，记录板 1 块，榔头 1 把，木桩 6 个，测钎 2 支，计算器 1 个（自备）。

12.3　实验方法和步骤

12.3.1　控制点布设和放样数据计算

建筑物轴线和高程测设前应先布设控制点。如图 12-1 所示，在空旷地面上，先打下一个木桩作为点 A，桩顶画十字线，以交点为中心，用钢尺丈量一段 50.000 m 的距离确定点 B（打木桩，桩顶画十字线）。本实验假设点 A 的坐标为 $x_A = 100.000$ m，$y_A = 100.000$ m；点 B 的坐标为 $x_B = 100.000$ m，$y_B = 150.000$ m。

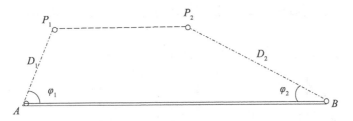

图 12-1　建筑物轴线测设

设点 A 的高程为 10.000 m，假设以上数据为已有控制点的已知数据。设建筑物的某轴线点 P_1 的坐标和高程为 $x_{P_1} = 108.360$ m，$y_{P_1} = 105.240$ m，$H_{P_1} = 10.150$ m；点 P_2 的坐标和高程为 $x_{P_2} = 108.360$ m，$y_{P_2} = 125.240$ m，$H_{P_2} = 10.150$ m。在实际实验中，根据指导老师的要求拟定相关数据；在实际工程中，根据设计图纸上的数据进行放样。

在控制点 A 和点 B，用极坐标法测设轴线点 P_1 和点 P_2 的平面位置，用水准仪放样其高程。

用极坐标法测设点 P_1 时，有

水平距离　　　　　　　　　$D_1 = \sqrt{(x_{P_1}-x_A)^2+(y_{P_1}-y_A)^2}$

与已知方向之间的水平夹角 $\varphi_1 = \arctan\dfrac{y_B - y_A}{x_B - x_A} - \arctan\dfrac{y_{P_1} - y_A}{x_{P_1} - x_A}$

同法可计算测设点 P_2 的数据 D_2 和 φ_2。

在表 12-1 中计算所需数据，并画出建筑物轴线测设略图。

12.3.2 极坐标法轴线点平面位置测设

（1）将全站仪安置于点 A，瞄准点 B，变换水平度盘位置使读数为 $0°00'00''$，逆时针旋转照准部，使水平度盘读数为 $360°-\varphi_1$，用测钎在地面标出该方向，在该方向上距离点 A 的水平距离为 D_1 处打下木桩；再重新用全站仪标定方向和用钢尺量距，在木桩上定出点 P_1。

（2）将全站仪安置于点 B，用同样的方法测设点 P_2（不同之处为瞄准点 A 后，照准部顺时针旋转 φ_2 角）。

（3）根据 P_1，P_2 两点的设计坐标计算两点间的水平距离，用钢尺丈量点 P_1 和点 P_2 之间的距离，测量值与计算的理论值之差不应大于 10 mm。

12.3.3 直角坐标法放样轴线平面位置

放样方法参照实验 10 中全站仪坐标放样部分。

12.3.4 轴线点高程测设

如图 12-2 所示，将水准仪安置于与点 A、点 P_1 和点 P_2 大致等距离之处，点 A 处的木桩上立水准尺，读得后视读数 a，根据点 A 的高程 H_A 可求得水准仪的视线高程（仪器高程）H_I。

$$H_I = H_A + a$$

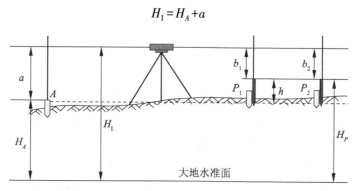

图 12-2 轴线点高程测设

点 A 的高程与点 P_1 和点 P_2 的设计高程之差（H_P 为待放样点的高程）为

$$h = H_P - H_A$$

点 P_1 和点 P_2 上水准尺应有的读数为

$$b = H_I - H_P$$

在点 P_1 和点 P_2 旁边各打一个木桩，用逐步打入土中的方法使立于其上的水准尺读数逐渐增大至 b，桩顶即为轴线点的设计高程。

12.4 注意事项

（1）小组成员独立计算测设数据，将计算结果相互进行校核，确定正确无误后再进行测设。

（2）将实验观测数据填入表 12-1 中，实验结束时，上交表 12-1。

表 12-1　测设数据计算记录

班级：_____　　组号：_____　　日期：_____　　观测者：_____

仪器编号：_____　　记录者：_____

平面测量放样数据记录					
测量点名称	坐标或坐标增量		水平距离/m	坐标方位角/ (°′″)	水平夹角/ (°′″)
	X（或 ΔX）	Y（或 ΔY）			

高程测量放样数据记录					
已知控制点高程/m	待放样点高程/m	后视点读数/m	视线高/m	待放样点读数计算值/m	备注

测设示意图

实验 13　道路圆曲线测设

13.1　实验目的

（1）掌握圆曲线主点元素的计算和主点的测设方法。
（2）掌握用偏角法进行圆曲线的详细测设。

13.2　实验安排和设备

（1）实验安排：实验课时为 3 学时；实验小组由 3 人组成。
（2）实验设备（以小组为单位）：全站仪 1 台，棱镜 1 个，测钎 10 支，木桩 3 个，榔头 1 把，记录板 1 块。

13.3　实验方法和步骤

13.3.1　圆曲线主点测设

道路圆曲线主点在测设之前，需要有标定路线方向的交点和转点。如图 13-1 所示，在空旷地面上打一个木桩作为路线交点 JD_1，然后向两个方向延伸 30 m 以上，确定出两个转点 ZD_1 和 ZD_2，并在转点处插上测钎。

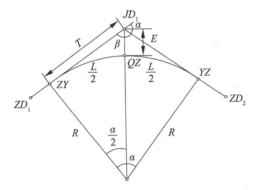

ZY—直圆点；QZ—圆曲线中点；YZ—圆直点；R—圆曲线半径；L—曲线长；
E—外距；T—切线长；β—转折角；α—路线偏角。

图 13-1　圆曲线的主点元素

在点 JD_1 安置全站仪，测定转折角 β，计算路线偏角 α（$\alpha=180°-\beta$）。按下列公式计算圆曲线元素（切线长 T、曲线长 L、外距 E、切曲差 J），并将结果记录于表 13-1 中。

$$T = R\tan\frac{\alpha}{2} \tag{13-1}$$

$$L = R\alpha \frac{\pi}{180°} \tag{13-2}$$

$$E = \frac{R}{\cos\frac{\alpha}{2}} - R \tag{13-3}$$

$$J = 2T - L \tag{13-4}$$

用安置于点 JD_1 的全站仪先后瞄准点 ZD_1 和点 ZD_2 以确定路线方向，用钢尺在该方向上测设切线长 T，定出圆曲线的起点 ZY 和圆曲线的终点 YZ，并打下木桩，重新测设一次，在木桩顶上标出点 ZY 和点 YZ 的精确位置。

用全站仪瞄准 YZ，将水平度盘的读数置于 $0°00'00''$，照准部旋转 $\beta/2$，确定转折角分角线的方向，用钢尺测设外距 E，定出圆曲线中点 QZ。

13.3.2 主点桩号计算

位于道路中线上的曲线主点桩号由交点的桩号推算而得。

$$\begin{cases} ZY \text{桩号} = JD_1 \text{桩号} - T \\ QZ \text{桩号} = ZY \text{桩号} + L/2 \\ YZ \text{桩号} = QZ \text{桩号} + L/2 \end{cases} \tag{13-5}$$

为了避免计算错误，可用下式进行检核，如果两次算得的 YZ 桩号相等，则证明计算正确。

$$YZ \text{桩号} = JD_1 \text{桩号} + T - J \tag{13-6}$$

13.3.3 用偏角法详细测设圆曲线

设圆曲线上每 10 m 需要测设里程桩，则 $l_0 = 10$ m，l_1 为曲线上第一个整 10 m 桩 P_1 与圆曲线起点 ZY 间的弧长，如图 13-2 所示。

用偏角法详细测设圆曲线，按下式计算测设点 P_1 的偏角 Δ_1 和以后每增加 10 m 弧长的各点的偏角量 Δ_0：

$$\Delta_1 = \frac{l_1}{2R} \times \frac{180°}{\pi} \tag{13-7}$$

$$\Delta_0 = \frac{l_0}{2R} \times \frac{180°}{\pi} \tag{13-8}$$

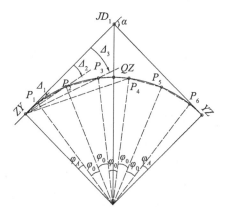

图 13-2 偏角法测设圆曲线

P_2，P_3，$\cdots$，P_i 等细部点的偏角按下式计算：

$$\Delta_2 = \Delta_1 + \Delta_0 \tag{13-9}$$

$$\Delta_3 = \Delta_1 + 2\Delta_0 \tag{13-10}$$

......

$$\Delta_i = \Delta_1 + (i-1)\Delta_0 \tag{13-11}$$

曲线起点至曲线上任一细部点 P_i 的弦长 C_i 按下式计算：

$$C_i = 2R\sin\Delta_i \tag{13-12}$$

曲线上相邻整桩间的弦长 C_0 按下式计算：

$$C_0 = 2R\sin\Delta_0 \tag{13-13}$$

曲线上任两点间的曲线长 L 与弦长 C 之差（弦弧差）按下式计算：

$$L - C = \delta = \frac{l^3}{24R^2} \tag{13-14}$$

偏角法详细测设圆曲线的步骤如下：

（1）将全站仪安置于点 ZY，照准 JD_1，变换水平度盘的位置，使其读数为 $0°00'00''$。

（2）顺时针方向转动照准部，使水平度盘的读数为 Δ_1，在全站仪所指方向上从点 ZY 用钢尺测设 C_1，得到点 P_1 的位置，用测钎标出。

（3）再顺时针方向转动照准部，使水平度盘的读数为 Δ_2，从点 P_1 用钢尺测设弦长 C_0，与全站仪所指方向相交，得到点 P_2 的位置，也用测钎标出，依此类推，测设各桩。

（4）曲线总长较短时，也可根据全站仪所指偏角方向，从曲线起点量弦长 C_i，得到点 P_i 的位置。

（5）YZ 的偏角应等于 $\alpha/2$，从曲线上最后一点至点 YZ 的距离应等于其计算的弦长。如果二者不符合，其闭合差应满足如下规定：半径方向（横向）的闭合差不超过 ± 0.1 m；切线方向（纵向）的闭合差不超过 $\pm\dfrac{L}{1\,000}$。

13.4 注意事项

（1）圆曲线主点测设元素和偏角法测设数据应经过两人独立计算，结果校核无误后再进行测设。

（2）本实验所占场地较大，仪器工具较多，实验结束后应及时查收，防止遗失。

（3）实验结束后，应上交表 13-1。

表 13-1　圆曲线测量记录表

班级：_____　　组号：_____　　日期：_____　　观测者：_____
仪器编号：_____　　记录者：_____

圆曲线元素名称	数值	圆曲线元素名称	数值
交点 JD_1		切线长 T	
转折角 β		曲线长 L	
偏角 α		外距 E	
圆曲线半径 R		切曲差 J	

圆曲线里程桩号	相邻桩点弧长 l/m	偏角 $\Delta/(° ′ ″)$	弦长 C/m	相邻桩点弦长 C_0/m

实验 14　无人机地形图测绘

14.1　实验目的

（1）了解无人机测绘仪器的组成。
（2）熟悉无人机的操作方法，掌握外业数据采集以及三维测图模型的生成方法。

14.2　实验安排和设备

（1）实验安排：实验课时为 2 学时；实验小组由 5~6 人组成，本实验为观摩演示性实验。
（2）实验设备（以小组为单位）：无人机摄影测量设备一套（含电池、摄像机、遥控器等）。

14.3　实验方法和步骤

14.3.1　准备工作

航测作业前，应充分收集与测区有关的地形图、影像等资料，了解测区的地形地貌、气候条件以及机场、重要设施等情况并进行分析研究，尤其应注意起飞点是否为禁飞区。

14.3.2　现场勘查

（1）了解测区的位置、地物和地貌分布情况、通视条件、道路交通及人文、气象、居民地分布等情况，并根据收集到的控制点资料，找到测量控制点的实地位置，确定控制点的可靠性和可使用性。
（2）检查测区的网络 RTK 覆盖情况，确定信号强度和可使用性。需要注意的是，拍摄应该在光线充足的天气进行。起飞点应选择空旷无遮挡的场地，这样可以获得良好的 GNSS 信号、通信距离、网络信号及开阔的视野。起飞点及飞行路线应远离 Wi-Fi 网络覆盖区域、信号塔、变电站等，以免无人机通信链路被干扰，导致飞行器失控及 RTK 失锁。
（3）布置像控点。像控点是在影像上能够清楚地辨别，且具有明显特征和地理坐标的地面标识点。像控点的地理坐标可以通过 GPS、RTK、全站仪等测量技术获取。

14.3.3　外业数据采集

1. 仪器准备

安装无人机螺旋桨，安装电池（确保无人机电池和遥控器电池的电量满足作业要求），检查 SD 卡，检查视觉系统的摄像头是否清晰无污点，检查螺旋桨是否安装正确，检查机臂、脚架套筒是否紧固。

2. 任务规划

外业工作中，无人机手动飞行的难度大，需要借助于航线规划软件来完成，比如使用大疆的航线规划软件 DJI GO 和 DJI FLY，还可以使用 Lichi 和 Smart Flight 进行航线规划。执行任务前，首先设置规划参数，包含作业区域和作业类型。其中，作业区域是指需要拍摄的区域；作业类型是指建模的类型，例如，是二维建模还是三维建模。

3. 参数设置

飞行参数设置包括飞行高度（100 m 左右，根据测区内障碍物的高度确定）、重叠率（70%~80%）、飞行速度、航线角度（主航线角度通常为 90°）等。

使用飞行器 RTK（已有千寻 CORS）的定位功能时应注意，当 RTK 模块异常时，手动关闭飞行器 RTK 的定位功能，并切换回 GNSS（global navigation satellite system，全球导航卫星系统）模式。飞行器起飞后再开启 RTK 的定位功能，导航系统将继续使用 GNSS 模式。

返航点设置：当 GNSS 信号首次达到四格及以上时，将记录飞行器的当前位置为返航点。后续也可以更新返航点，比如：以飞行器当前位置为返航点，或者以遥控器当前位置为返航点。返航高度要高于测区内最高建筑物的高度。

报警电压设置：若飞行器返航时为逆风飞行或电池循环次数较多（超过 100 次），则建议把低电量报警电压的阈值增大。

航线规划：不同的航线规划软件的规划方式略有差异，下面以 DJI GO 和 DJI FLY App 为例示范。

DJI GO App 提供了五向飞行、井字飞行两种三维建模航线规划方式，五向飞行为 5 组飞行航线，包含 1 组正射航线和 4 组不同朝向的倾斜航线。其中，井字飞行航线由两条摄影测量 2D 航线垂直排列组成，系统默认云台的角度为 $-60°$。其中推荐旁向重叠率为 80%，航向重叠率为 80%。DJI FLY App 中的倾斜摄影为五向飞行。

在遥控器端主页面上点击"规划"按钮，选择作业模式后，点击地图上的点进行作业区域的规划，也可以借助于 91 卫图等软件生成测区的 KML 文件，把 KML 文件导出到电脑合适的位置，然后通过 USB 把电脑上的 KML 文件导入飞行器的遥控器存储中即可。

4. 任务执行

软硬件（无人机、螺旋桨、遥控器、电池、SD 卡读卡器等）检查无误后，先开启遥控器电源，再开启无人机电源。操控员在飞行器后方操作，先让飞行器在低空悬停 1 min 左右，使电池充分预热，以确保各类传感器正常工作。观察飞行器的飞行姿态是否正常，确认无误后，执行飞行任务，无人机开始自动执行航线任务。

执行任务过程中，观察回传画面及相机参数，当画面中出现过曝或欠曝时需调整相机参数。当无人机因电量不足或其他原因导致单架次无法完成航线任务时，系统会自动记录航线中断点。再次执行任务时需选取当前任务，点击"调用"按钮即可继续执行任务。

5. 外业成果检查

查看拍摄的照片，检查照片有无模糊、曝光不正常等不合格现象，并将照片进行备份。如有漏拍、模糊等异常情况，务必及时补拍。产生漏拍的原因可能有：选择的拍摄模式为"定距拍摄"；SD 卡读写速度慢或 SD 卡已满。

将采集完的数据导入成图软件，并进行空三处理。空三处理是指摄影测量中利用影像与所摄目标之间的空间几何关系，通过影像点与所摄物体之间的对应关系计算出相机成像时刻相机位置姿态及所摄目标的稀疏点云的过程。进行空三处理后，能快速判断原始数据的质量是否满足项目交付要求以及是否需要增删影像。二维重建和三维重建前都必须先进行空三处理。完成空三处理后查看空三报告，如果空三结果不满足要求，则需要重新飞行，采集数据，再进行空三处理，直至得到满足要求的空三结果。

6. 飞行器的贮存保养

任务完成后应及时擦拭飞行器，并将其存放在安全保护箱内，保持环境干燥。飞行器应定期保养。

14.3.4　内业数据处理

内业数据的处理流程如图 14-1 所示。图中，POS 数据即记录无人机拍照瞬间的三维坐标（经度、纬度和飞行高度）及飞行姿态（航向角、俯仰角和翻滚角）的数据；DSM（digital surface model，数字地表模型）是指包含了地表建筑物、桥梁和树木等高度的地面高程模型；DOM 表示数字正射影像（digital orthophoto map）。

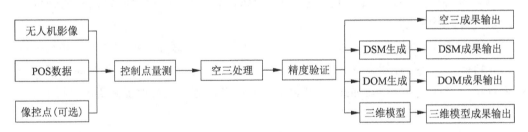

图 14-1　内业数据处理流程

1. 控制点量测

摄影测量任务完成后，取出相机 SD 卡，把 SD 卡中的数据转存入电脑。通过成图软件新建重建任务，添加无人机拍摄的照片，导入相机参数和像控点。像控点分为控制点和检查点，控制点用于优化空三的精度，可提升模型精度，也可实现地方坐标系或 85 高程系统的转换。检查点用于检查空三的精度，可通过检查点来定量对精度做评价。

2. 空中三角测量数据处理

首先通过软件刺像控点的方式将像控点与拍摄到该点的照片关联起来，然后通过软件进行空三数据处理。

3. 精度验证

设置重建参数，例如，选择重建分辨率、建图场景等。首先选择坐标系统，开始重建，建图完成后，地图界面显示重建结果，得到二维/三维的地形图。然后导出建图报告，通过报告可以直观地了解重建的质量。

4. 成果输出

大疆智图生成的正射影像和实景三维模型均可导入 CASS-3D 作数字线划图生产。数字线划图是与现有线划基本一致的各地图要素的矢量数据集，可保存各要素间的空间关系和相关的属性信息。

基于 DOM+DSM 测图的步骤如下：

（1）数据准备：在大疆智图二维重建的工程文件夹中，找到 DOM（result.tif）和 DSM（dsm.tif）成果及其配套的.prj 和.tfw 文件。

（2）模型转换：打开 CASS-3D 软件，点击"Build Dsm"按钮，在弹出的对话框中分别输入 DOM、DSM 和存储路径，并选择合适的清晰度（数值越大，清晰度越高，耗时越久），点击"开始生成"按钮，在设置的路径下生成一个.osgb 文件。

（3）打开模型：点击"3D"按钮，打开上一步转换的.osgb 文件。此文件是由 DOM 套合 DSM 的垂直模型，可基于此模型开展线划图绘制等操作。

14.4 注意事项

（1）若负载镜头内部起雾，则可通过开机预热加速镜头水汽的消散。

（2）应时刻关注电量的变化。

（3）保持飞行姿态平稳，注意飞行环境的变化（避免在积雪反光强烈的地区，极寒、暴风雪天气等飞行）。

（4）现场作业时，班组作业人员应与所属飞行管制分区建立可靠的通信联络，进行飞行计划报备。飞行计划报备一般包括：飞行准备情况、当日预计作业时间，以及作业结束时间（当日飞行结束时通报）。具体通报时间和内容按空域批复函要求执行。

第二部分　土木工程测量实习

实习目的

土木工程测量实习是土木工程专业的一项重要的实践性教学环节。实习的目的是让学生巩固所学的理论知识，掌握测量仪器的基本操作技能，着重培养学生的动手能力和解决实际问题的能力，使学生掌握大比例尺地形图测绘的的基本方法。

实习内容与要求

大比例尺（1∶500）地形图的测绘

根据需要选择控制测量的区域，并在所选区域中划定尺寸为 150 m×150 m 的部分作为地形图测绘的区域，按比例尺的精度要求测绘比例尺为 1∶500 的地形图一份。

（1）平面控制测量

根据外业测量选点的原则，选择合适数量及位置的控制点组成闭合导线，绘制草图，并做好标记，以图根导线测量的精度要求测定并计算各控制点的坐标，作为测区内的平面控制。角度容许闭合差 $f_\beta = \pm 60''\sqrt{n}$（$n$ 为测站数），导线的相对误差容许值 $K \leq 1/2\,000$。

（2）高程控制测量

在选定的测区，根据外业测量选点的原则，布置适量的水准点，点位可以与平面控制测量相同。绘制草图，做好标记，组成闭合水准路线，按照图根水准测量的要求进行高程控制测量，采用两次仪高法或者双面尺法进行测站检核，测量闭合差 $f_{h容}$ 应满足如下要求：$f_{h容} = \pm 12\sqrt{n}$ mm（n 为测站数）或 $f_{h容} = \pm 20\sqrt{L}$ mm（L 为路线长度，单位为 km）。

（3）地形图测绘

利用已设置好的控制点进行碎部点测量和高程测量；在 A2 绘图纸上划定合适的绘图区域并绘制坐标方格网（方格的总尺寸为 300 mm×300 mm，每个方格的尺寸为 100 mm×100 mm）、展绘控制点及细部点，各地物严格采用图式符号表示；以 0.5 m 为基本等高距绘出等高线（部分区域起始点的坐标与高程可自己假定），完整绘制限定区域内 1∶500 大比例尺地形图。图上需有图名、比例尺、高程系统、坐标系统、图例等，标题栏中注明图名、比例尺和完成人。

（4）放样数据计算及放样操作

在所绘地形图内选定一个点，根据已有的两个控制点用极坐标法计算放样数据，并说明放样方法，或者根据地形图上的相关信息计算该选定点的坐标，并结合已有的两个控制点采用全站仪坐标放样的方法完成放样。

实习成果要求

（1）高程控制测量记录及计算表一份（A4 纸大小）；

（2）导线测量记录及计算表一份（A4 纸大小）；

（3）测绘地形图一份（需把网格内所有标志性物体绘出），图纸规格为 A2（带签字栏）；

（4）2~3 个特征点的放样数据计算过程及结果，并说明放样方法（每人一份，A4 纸大小）；

（5）个人实习日志（不少于 6 篇）和实习小结（每人一份），书写在 A4 纸上。

需要注意的是，每个小组的实习报告按照水准测量记录及计算表、导线测量记录及计算表、放样数据计算过程及结果、实习日志、实习小结的顺序排列，然后将上述内容加上封面和目录后装订。

实习时间安排

土木工程测量实习的总时间为 2 周。

实习组织

1. 仪器与工具

各小组配备：经纬仪 1 台或全站仪 1 台，水准仪 1 台，水准尺 2 支，3 m）钢尺 1 把，花杆 2~3 根，油漆 1 桶，测钎 5~8 支，记录纸（本）1~2 本（自备），计算表格。

2. 人员组织

以班级为单位分为若干组，每组 3~5 人，每组设组长、副组长各 1 人，负责全组实习的安排与管理，小组成员在实习中必须密切配合，团结互助，以便按时保质完成实习任务。

成绩评定

实习作为一门实践课程，成绩记入学生成绩册。评分内容包括实习表现、实习报告和地形图。具体如下：

1. 实习表现（20%）

水准仪、经纬仪（或全站仪）的基本操作技能；实习态度及遵守纪律、团结协作等表现。

2. 实习报告（50%）

每组上交的内业计算书、成果数据真实，计算正确，数据误差满足要求；实习日志和小结的内容真实，书写认真，每篇日志字数不少于 300 字，小结字数不少于 2 000 字。

3. 地形图（30%）

每组上交的地形图应描述准确、标志清晰，且须有本组所有成员的签名。

注意事项

（1）相邻控制点间要通视，用油漆做好标记（有本组标志）并命名点号，且不要布置在路中间，以免被破坏。

（2）原始测量数据要记录准确、清晰。

（3）绘制方格网，保持图纸整洁，内容完整，表达规范。

（4）注意人身及财产安全。

参考文献

[1] SCHULTZ R J. Education in Surveying: Fundamentals of Surveying Exam [J]. Professional Surveyor, 2006, 26 (3): 38.

[2] GHILANI C D. Adjustment Computations: Spatial Data Analysis [M]. 6th ed. New York: Wiley, 2010.

[3] SCHOFIELD W, BREACH M. Engineering Surveying [M]. 6th ed. Burlington: Elsevier Ltd., 2007.

[4] ERNST C M. Direct Reflex vs. Standard Prism Measurements [J]. The American Surveyor, 2009, 6 (4): 48.

[5] HOFMANN-WELLENHOF B, LICHTENEGGER H, COLLINS J. GPS: Theory and Practice [M]. 5th ed. New York: Springer, 2001.

[6] ARMH A E, MOHAMED G E. Using Commerical-Grade Digital Camera Images in the Estimation of Hydraulic Flume Bed Changes: Case Study [J]. Surveying and Land Information Science, 2008, 68 (1): 35-45.

[7] CHILANI C, WOLF P R. Elementary Surveying: An Introduction to Geomatics [M]. 13th ed. Upper Saddle River: Prentice Hall, 2011.

[8] 陈学平. 工程测量 [M]. 北京: 中国建材工业出版社, 2014.

[9]《英汉测绘词汇》编辑组. 英汉测绘词汇 [M]. 北京: 测绘出版社, 1982.

[10] 顾孝烈, 鲍峰, 程效军. 测量学实验 [M]. 2版. 上海: 同济大学出版社, 2010.

[11] 中国有色金属工业协会. 工程测量标准: GB 50026—2020 [S]. 北京: 中国计划出版社, 2020.